The Vital Spark

AN INTRODUCTION TO THE GENERATION,
SUPPLY, AND USE OF ELECTRICITY

John P. Wright, BA, MSc, CEng, MIEE, MInstP
Senior Lecturer in Physics and Physical Electronics
Newcastle upon Tyne Polytechnic

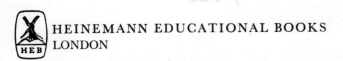
HEINEMANN EDUCATIONAL BOOKS
LONDON

Heinemann Educational Books Ltd

LONDON EDINBURGH MELBOURNE AUCKLAND TORONTO
HONG KONG SINGAPORE KUALA LUMPUR
IBADAN NAIROBI JOHANNESBURG NEW DELHI

ISBN 0 435 68945 2

© John P. Wright 1974

First published 1974

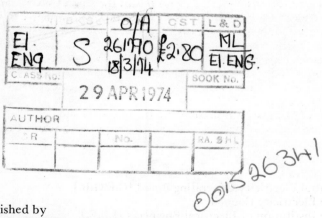
Published by
Heinemann Educational Books Ltd
48 Charles Street, London W1X 8AH

Set in 10 pt. IBM Baskerville, printed by photolithography,
and bound in Great Britain at The Pitman Press, Bath

Preface

This book is an attempt to provide a much-needed background reader in heavy-current electrical engineering. In my opinion there is as much — if not more — promise and excitement in this field than in any other branch of modern technology. I hope that in the book I have been able to communicate and transfer some of my enthusiasm for the subject.

To place the industry in perspective an introductory guide is provided on the history of electrical engineering. One of the main objectives of the book is to illustrate the wide range of activities of a professional electrical engineer. He must be something of a scientist, sociologist, economist, and usually a manager as well. To demonstrate these aspects particular sections are developed in detail e.g. the physics of generators, industrial design in the control room and the selection of materials. In this respect it is hoped the book will be of value to sixth formers. Perhaps it may even provide a guide to a future career!

The book does not set out to provide new material. The source information was already largely available in journals. I have attempted to collect all that seemed relevant and important to me and which I think might be of interest and value to others. No writer can be sure of the origin of all the facts. Wherever I have consciously borrowed I have given grateful acknowledgement. Where an acknowledgement has not been made where it was due, I would ask those concerned for their charity.

Special thanks are offered to Mr R. D. Harrison of the Open University, who suggested the idea of the book and to Mr G. B. Jackson, formerly of the C.E.G.B., for his editorial help and in providing material for the chapter on 'Transmission and Distribution'. Thanks are extended to Mr C. MacKechnie Jarvis for his invaluable articles on the history of electrical engineering in the I.E.E. Journal, to Mr B. Bowers and Mr A. Ridding for their useful Science Museum booklets and to Dr R. Hawley of Reyrolle—Parsons Ltd., for his help in preparing the chapter on 'Generation'.

I wish to acknowledge the help supplied by the technical and public relation staffs of the following firms and organizations without which this work would have been out of date before it started.

> Reyrolle—Parsons Co. Ltd.
> G.E.C. Ltd.
> Central Electricity Generating Board [C.E.G.B.]
> The Electricity Council
> The Institution of Electrical Engineers [I.E.E.]
> The Science Museum

Finally, sincere thanks are given to my family for their direct and indirect help.

1973   J.P.W.

Contents

1 Electricity Today

The prosperity of any nation nowadays can be judged by the amount of
electricity it uses and the growth of that demand. Electricity is definitely one

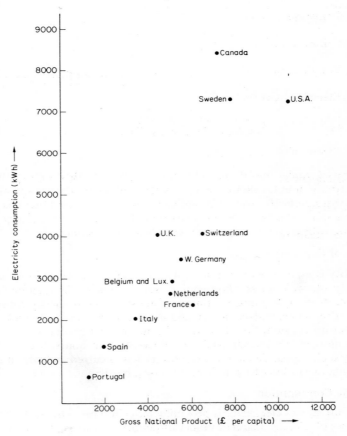

Electricity consumption (kWh) and gross national product (£)
per capita 1968.
(Source: U.N. World Energy Supplies – Series J, No. 13.,
U.N. Monthly Bulletin of Statistics.)

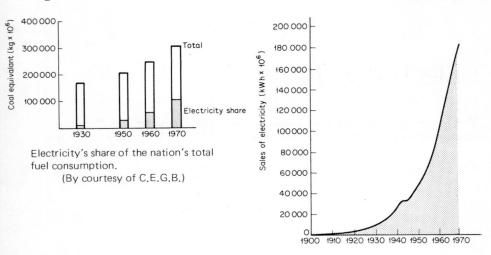

Electricity's share of the nation's total
fuel consumption.
(By courtesy of C.E.G.B.)

Consumption of electricity in England
and Wales. (By courtesy of C.E.G.B.)

of the very foundations of our modern civilization. In this country the demand
for electricity doubles about every ten years and, as an illustration, applying
this growth rate to the average home means that there could be twice as many
electrical appliances in use in ten years time. In fact, however, this growth is
spread over domestic, commercial and industrial consumption.

Variety of Consumers

Electricity is used by a wide variety of consumers, from large factories making
cars or textiles to domestic households using it to drive the vacuum cleaner
and cook the Sunday joint. Fortunately industrial and domestic consumption
are about the same, and as the demand in the house tends to increase at times
when factory consumption decreases, this means that generating stations can
be kept more evenly loaded over the full twenty-four hours. It is interesting to
ask why such a great demand for electricity has occurred and why it goes on
increasing. Let us look first at domestic consumption.

Domestic Consumption

As time progresses electricity is providing more and more of the basic
necessities of life, i.e. a larger number of people are turning to electricity for
heating, cooking, lighting and cleaning in the home. Secondly, with increased
standards of living the general public are demanding more and more luxury
goods, many of which require electric power, such as record players, radio and
television receivers, central heating installations and air conditioning plants.

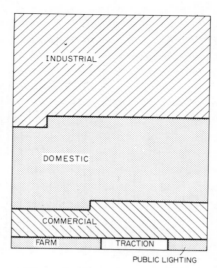

The types of electricity consumer.

Consider the example of heating. It is obvious why electricity should have advantages over other fuels such as coal or oil. It is clean and safe, needs no storing, and provides an instantly available 'fuel' at the flick of a switch. The Clean Air Act of 1956 and the strengthening measure in 1968 made it illegal to burn smoke-emitting fuels (in smokeless zones) and electricity has profited from this. The advent of cheap 'off-peak' electricity to supply domestic space heating has further increased its popularity.

Off-peak heating with block storage heaters takes the form of warming a block of artificial 'stone' with about 2–3 kilowatt (kW) of power during the night and allowing the stored heat to dissipate into the room during the rest of the day. In some cases the heating element is buried in a concrete floor which acts as the heat store.

Industrial Consumption

Industrial demand has also increased at a very rapid rate. There are four main reasons:

1. In the last few decades our standard of living has increased so that there has been a great demand for capital and consumer goods, with a consequent need for higher productivity. Work in factories was therefore required to be done more quickly. Electricity has increased productivity by enabling more power drives and automatic processes to be introduced.

2. The cost of manpower today, especially skilled manpower, is very high. Manufacturers must of necessity, therefore, get most out of their available

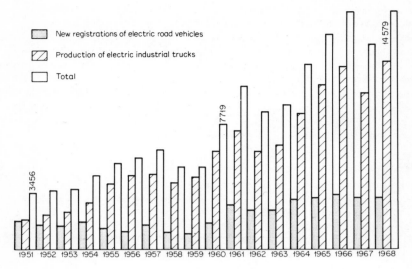

Growth of electric vehicle production.
(By courtesy of the Electric Vehicle Association of Great Britain.)

manpower by providing as much automation as possible to increase the
'power per worker'. Power per worker is yet another yardstick with
which to measure the advance of technology.

3. Some jobs are pleasant and interesting, others unpleasant, dull and
   monotonous. Electricity can reduce the number of uninteresting jobs by
   operating mechanized tools which enable the routine work to be completed quickly. There is a genuine humanitarian motive involved too, in
   reducing the number of arduous and dangerous jobs, for example in the
   coal mining industry. In former times coal was hacked from the coal face
   by men working with picks and shovels in the most filthy conditions
   imaginable, but today electrically powered machines cut away the coal
   and feed it on to a transport conveyor belt. In the latest coalmines,
   electric power, backed by very advanced automatic control, has taken
   over nearly every repetitive task previously done by manual labour.

4. Finally, some jobs are best done by a machine, since once the machine
   has been set up correctly it can produce consistently more accurate work
   than can manual methods.

Commercial Consumption

Commercial uses for and users of electricity are increasing as well as the
domestic and industrial demand. Shops, offices and hotels are all using more

electricity for lighting, refrigeration, space and water heating. A still higher demand was created by the Offices, Shops and Railways Premises Act of 1963, which aimed at bringing standards of health and safety in these premises into line with those long enforced in factories. It made compulsory adequate lighting, heating, hot water supply and clothes drying facilities for employees.

Most office machinery uses electricity. Magnetic tape recorders, electronic desk calculators and photocopying machines are a few examples. Communication between offices and firms is facilitated by telephone and telex systems. On-line computer terminals enable office workers to have immediate access to information storage and data processing facilities.

Railways

A major electrification scheme is now in full service from London to Birmingham, Liverpool and Manchester. Locomotives of some 2500 kilowatt haul both passenger and freight trains at higher speeds and with better acceleration than could the old steam engines. This increases the density of traffic between the cities and allows the line to carry more traffic. (By 1970 the modernization was such a success that electrification was approved for extension to Glasgow.)

Farming

It was announced recently that well over 98 per cent of farms in this country have been connected to the public supply system. This implies that not only can the majority of the farming community now enjoy the benefits formerly available only to urban dwellers, but they can also use electricity to industrialize their farms. Electricity can be used in the preparation of animal foodstuffs, for grain-drying and to provide heat and light for the rearing of calves, pigs and poultry. Electric milking machines are now standard equipment on dairy farms.

Horticulture

In horticulture electricity can be used to provide thermostatically controlled heating, ventilation, soil warming and sterilizing. In fact, with the aid of electricity it is now possible to provide home-grown flowers all the year round. Chrysanthemums, for example, can be grown for any time of the year because it is possible by artificial lighting to control the 'daylight' hours during the winter months.

Demand for Electricity

Let us look now at the way in which the total national demand for electricity varies throughout twenty-four hours, see graph, page 7. The size and shape of the demand graph varies between summer and winter. Notice that even with the domestic and cheap off-peak heating loads there is still a large variation in total demand. Perhaps you can account for the peaks and troughs of the curves?

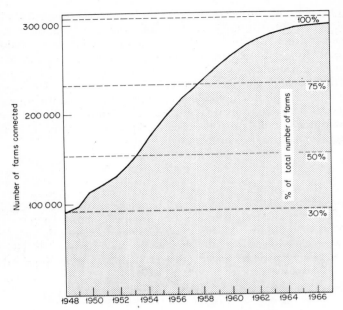

Number of farms connected to the National Grid.
(By courtesy of Electricity Council.)

An engineer at the National Control Centre has to be a bit of a sociologist, for the meters indicating various power flows at different times of the day reflect the social patterns of our daily life. Demand is high during the day when we are working and low at night when most of us are asleep. In winter the evening load is higher than in summer because the heating, cooking and lighting load in the home, plus the switching-on of street and shop lighting, overlaps the later afternoon load when factories are still running at full capacity.

In view of the way in which the demand for electricity has increased, is it reasonable to expect the demand to continue to rise? From the evidence available there seems to be no reason to doubt that it will rise for some time to come, although it may not always grow at the same rate. There will be minor fluctuations in the growth rate since this is linked with the nation's general economic progress. One of the most important tasks in the electricity power supply industry is to predict what the demand for electricity is going to be in a few years time so that power plant can be built to meet the demand. The need to plan so far ahead arises from the fact that it takes five years to build the giant power stations to produce the power. 'Giants' are necessary because it is cheaper to construct and install a few large-sized plants than a large number of small-sized ones. That is to say, the cost per kW installed falls as the size of generators and boilers increases.

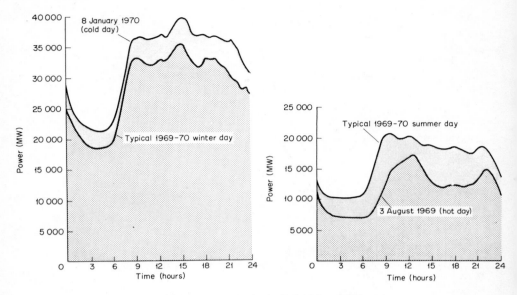

Summer and winter demands for electricity.
(By courtesy of C.E.G.B.)

Modern Power Stations

Suppose you were given the job of deciding how many and what type of power stations to build in the next ten years. How would you start? What factors would you think important in helping you to make a decision? Here are a few suggestions. You would need to know what plant capacity is at present available, and how much of that plant will be serviceable in ten years time. You will need to know what is the present demand for electricity, and the most probable demand ten years hence. What about spare plant? How much extra has to be allowed for unforeseen breakdowns? These are the sort of questions one has to answer before a decision can be made on how much more plant is needed. You also have to consider where stations should be built based on the loading of the system, availability of fuel, the supply of cooling water, etc.

What about the types of power station? Do we build all coal-fired stations or all nuclear-powered stations? Perhaps we should build a few of each. Why? If we decide to do so what is to be the proportion of each type and where shall they be situated? What about oil-fired, diesel, hydro-electric power and pumped storage stations? Where do they come into the picture?

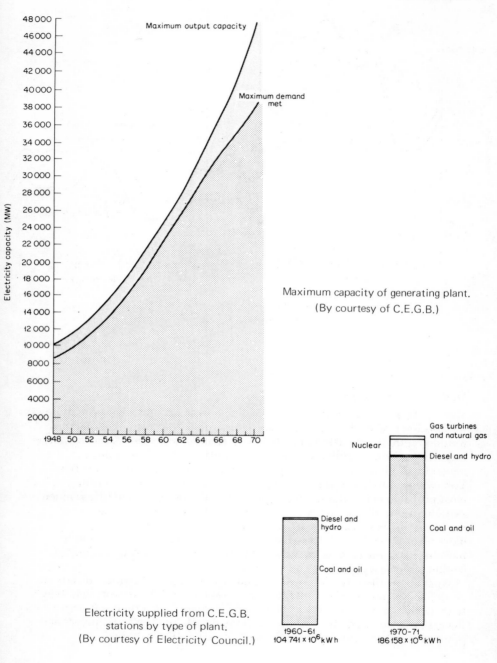

Maximum capacity of generating plant.
(By courtesy of C.E.G.B.)

Electricity supplied from C.E.G.B.
stations by type of plant.
(By courtesy of Electricity Council.)

The decision on the types and locations of power stations to be built is complex and difficult and involves some obvious and other less obvious factors. Let us look at a few of the factors that might influence our eventual selection.

Besides capital cost, running costs and security of supply, there will be political factors, social factors, and amenity values to be considered. Of one thing we can be certain, any final choice will involve some compromise and the engineers concerned will need to be sociologists as well as experts in their technology.

Types of Power Stations*

(i) Coal

Coal is the most important source of fuel used in power stations today; some 65 million tonnes‡ are consumed annually in England and Wales. With such large amounts it is obvious that transport of the coal from the mine to the station will be an expensive item. The fact that the cost of mining coal varies from one area to another must also be taken into account. To appreciate the size of the problem, if you could load all the coal burnt in British power stations in one year in rail wagons all at once, the train would stretch from London to South America and back again! To reduce the transport cost to a minimum, coal-fired stations are placed as near as possible to the cheapest sources of coal. Thus a large number of stations have recently been built on the East Midlands coal fields.

With such vast amounts of coal used daily, there is a considerable amount of waste ash — especially from the poorer coals — and transportation and dumping costs are considerable. Some stations have turned this liability into an asset. The ash is bought to make bricks and concrete, to build embankments under roads and to reclaim agricultural land from disused quarries. Ash has even been used in the repair of the massive nave pillars in Winchester Cathedral.

The last few years have seen a general recession in the coal mining industry and the government is concerned that the run-down should be gradual and controlled so as to cause as little hardship as possible. Thus the government would aim at keeping certain pits working for social rather than economic reasons. As a result of this policy there is often political pressure on the Generating Board to build coal-fired stations where their choice might have been otherwise.

(ii) Oil

Oil-fired power stations are usually built near an oil refinery. This is because the fuel is so 'heavy' and viscous that it has to be pumped from the refineries to the power station whilst still hot. ('Heavy' fuel oil is the residual oil left

*See Appendix I
‡1 tonne = 1000 kilogramme (kg).

High Marnham coal-fired power station in the East Midlands —
the first 1000 megawatt station to be built by the C.E.G.B.
Preparatory building work commenced in February 1956 and the
last of the five 200 megawatt turbo-generator sets was commissioned in June 1962.

Because of the location, cooling towers are used to cool the
circulating water needed to remove the heat from the steam in the
turbine condensers. Inside the cooling towers, the heated water from
the condensers is sprayed so that it forms a dense 'rain' which is
cooled by both evaporation and ascending air currents. The cooled
'rain' water is then collected and returned to the condenser for re-
use. The tall chimneys discharge 'cleaned' gases from the boilers
high into the atmosphere. Note the way the building has been
massed to obtain a reasonably satisfactory appearance.

(By courtesy of C.E.G.B.)

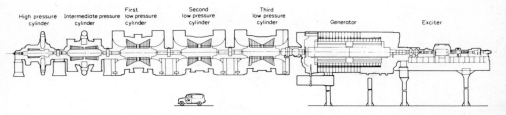

High pressure | Intermediate pressure | First low pressure cylinder | Second low pressure cylinder | Third low pressure cylinder | Generator | Exciter

Cross-section of 660 megawatt turbo-generator. The mini van
gives an idea of the size of the machine.
(By courtesy of C.E.G.B.)

behind after the lighter fractions have been distilled off in the refinery.) For
example, the power station at Fawley, Southampton, is built near the Esso
refinery.

Oil is a relatively cheap fuel but its use is limited by a very heavy tax and
the problem of reliability of supply. Most of our oil comes from the Middle
East and an unstable political situation in that area may easily result in the
cutting of supplies. The long pipe lines crossing the desert are vulnerable to
attack, sabotage and political bargaining. Large oil tankers have been constructed to bring oil from the Middle East via the Cape route but the journey
is long and therefore expensive. Supplies from other countries such as
Venezuela are also available, but the distances involved are still very great and
the cost therefore high. If oil resources are developed in Libya or exploited in
sufficiently large quantities under the North Sea, or if the Government
lowers the tax, then perhaps the building of oil-fired stations will become
more attractive. A point in their favour is that the fuel-handling plant is much
cheaper than for coal and there is no waste product (i.e. ash) to handle.

(iii) *Nuclear*

Nuclear stations are basically similar to coal and oil-fired stations except that
the heat is generated in a nuclear reactor rather than in a furnace. Because of
the radiation hazard, extra safety precautions have to be built into the design.
The capital cost of a nuclear station is at present about 50—100 per cent more
than that of an equivalent coal or oil station.

In this country nuclear stations fall into two categories — either the
Magnox or the *Advanced gas cooled* type.

The Magnox reactor uses natural uranium as a fuel; this is enclosed in
special cans made of magnesium alloy — hence the name 'Magnox'. Carbon
dioxide gas is blown past the cans to remove the heat produced internally,
and the hot gas is used to heat water to produce steam to drive the turbines.

The Advanced gas cooled reactor uses enriched uranium fuel and because
of this the reactor can be made much smaller and allowed to run hotter. As
before, hot carbon dioxide gas is used in the production of steam, but at higher
temperatures and pressures. Hence this type of reactor is more efficient than

Oldbury nuclear power station situated on the east bank of the
river Severn has twin Magnox reactors giving an electrical output
of 530 megawatt. 'Quick-start' gas turbine generators can boost
this figure to over 600 megawatt.

An important factor in the choice of site was the existence of
a rock shelf in the river which facilitated the construction of a
reservoir of area 1.5 square kilometre from which the station
draws its cooling water. The view shows, left to right, part of the
cooling water system, the reactor block which contains two
reactors and is linked by a common services building, the central
control building, the three gas turbine chimneys and the end of
the turbine hall.

(By courtesy of C.E.G.B.)

Table 1

Nuclear and Coal-fired Power Stations: Steam Temperature
and Pressure Conditions

Generating Station	Turbine Inlet Conditions	
	Steam Temperature ($^\circ$C)	Steam Pressure (bars)*
Nuclear (Magnox) e.g. Dungeness 'A' — 142 megawatt sets	391	38
Nuclear (A.G.R.) e.g. Dungeness 'B' — 660 megawatt sets	566	160
Coal-fired e.g. Drax — 660 megawatt sets	566	160

*N.B. 1 bar = 10^5 N m^{-2} = 1 atmosphere of pressure

the Magnox reactor. Conventional steam plant, of the kind employed in coal-fired stations, is used.

In both types of reactor little nuclear fuel is required. About 1 kilogramme of uranium gives the same output of heat as 56 000 kilogramme of coal. Nuclear power stations can be situated in areas having little relation to the fuel source and its cost of transport. They were previously sited in areas of low population density because of the hazard of radiation, but the safety record is so good that future stations will be sited closer to towns; it was decided in 1969 that the new nuclear power stations at Hartlepool and Heysham would be near urban areas.

Another possible source of hazard with nuclear stations lies in the disposal of radioactive waste. The fission products in a nuclear reactor are highly radioactive and so, after processing to recover certain materials, the waste products are stored in special containers in selected underground sites, which are controlled by the Atomic Energy Authority, until the natural processes of radioactive decay reduce the danger. Nuclear stations have a high capital cost but a low running cost due to the small amount of fuel required; therefore they form the base load stations of the National Grid, and are kept in constant operation all the year round.

Other types of nuclear power plant are at present being developed. Perhaps the most promising is the Prototype Fast Reactor (P.F.R.) at Dounreay in Scotland; this both consumes plutonium and produces more of it from 'natural' uranium, and has an electrical output of 250 megawatt (MW). It is envisaged that this will be the dominant type of reactor installed in the large commercial power stations in the 1980s and onwards.

These three types of power station, coal, oil and nuclear, all need a plentiful supply of cooling water to enable the steam turbine to work efficiently. A big modern power station could require up to 250 000 cubic metres of water per hour. It is desirable, therefore, that the stations be sited near a river, a lake, an estuary or the sea. In these cases the cooling water can be returned directly to the source after use and cooling towers are not required.

On the other hand, power stations situated on smaller rivers do not have un-limited water supplies so that cooling water is passed around a closed-circuit and through cooling towers (where its heat is removed by evaporation) and re-used. Where cooling towers are employed, about one hundredth part of the cooling water evaporates and has to be made up. It is this make-up water which is supplied by the small river. At the same time the river water is also used for purging the cooling tower ponds to keep salt levels low. This latter process 'aerates' the outgoing water and helps to keep the river 'clean' by raising the oxygen content.

Water discharged from power stations is warm and most river authorities require the water to be cooled to a certain temperature before being returned to the river. Thermal pollution of a river can have a most damaging effect on its natural life.

Considerable attention has been given to ways of making use of warm water discharged from direct-cooled power stations. District heating has been studied but has so far proved uneconomic. Another project, fish farming, has however shown encouraging results in the experimental stages. Certain types of fish, prawns for instance, flourish when reared in special tanks of warm water. There is every hope that this technique may before long contribute significantly to the nation's food supply.

(iv) Gas Turbines

Twin gas turbines driving generators up to 60 MW have recently been introduced to help out the National Grid in emergencies and at peak-load times. The turbines are based on an aircraft-type jet engine which can be started up and brought on load very quickly. The 'cold start' can be done very quickly with-out the need for external power to feed the auxiliary plant during running up. This type of 'prime mover' might assume considerable importance if the natural gas recently discovered in the North Sea turns out to be a suitable fuel for gas turbines.

In 1967 a big boiler at Hams Hall 'C' power station in the Midlands was converted to natural gas for an experimental period. The results of the trial were so encouraging that in 1970 the Board applied and were granted permis-sion to extend the scheme to the other five boilers in the station. When the scheme is fully operational the C.E.G.B. will be the second largest consumer of gas in the U.K. — the largest is I.C.I. — and the 390 MW station will use up to 30 000 million million joules per year. The extent to which gas will be used as a primary fuel for electricity generation still remains in doubt, mainly because of incomplete knowledge about the extent of reserves and the fact that the C.E.G.B. has to purchase gas through the Gas Council — their biggest commercial rivals.

(v) *Hydro-Electric*

Hydro-electricity is only feasible in wet, mountainous areas such as Wales and
Scotland, since a large water catchment area is necessary with a high, prefer-
ably regular, rain-fall. A hydro-electric scheme consists essentially of a dam
forming a reservoir, pipes leading the water to a lower level, a water turbine
and an electrical generator. Often hydro-electric power stations are linked
with water control and irrigation schemes.

At present it is not possible to store electricity generated at night, when
demand is low, for use at peak demand periods during the day. However,
hydro-electric pumped-storage schemes provide a means of solving this problem
to some extent. Electricity taken from the Grid at night — when it is cheap —
is used to pump water to a high dam — hence electrical energy is effectively
'stored' as potential energy given to the water. During the day, when the
demand for power reaches a peak, the stored water is allowed to run down to
a water turbine and a generator. The generated electricity is very valuable to
the hard-pressed Grid. Pumped-storage schemes usually operate for only a
few hours per day, but can reach full load in a few minutes.

Power Stations and Costs

A typical modern coal-fired power station would consist of four generators,
each having a power of 500 million watt or 500 megawatt, giving a total
capacity of 2000 megawatt. The capital cost of installation, i.e. plant purchase
price or cost of construction, would be about £100 million — £50 per kilowatt.
For a nuclear station the capital cost would be about £90 per kilowatt,

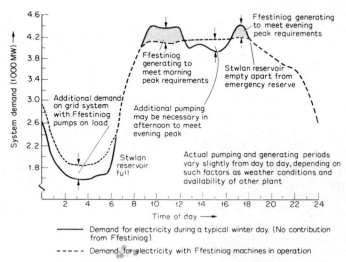

Typical winter load curve with operation of Ffestiniog superimposed.
(By courtesy of G.E.C. Machines Ltd.)

Ffestiniog pumped storage scheme, designed to supply electricity to the National Grid at peak demand periods and to act as reserve generating capacity in case of abnormal plant breakdown elsewhere.

The plant comprises four vertical water pump/turbine — motor/generator sets with a maximum output of 360 megawatt. From midnight to about 6 a.m. water is pumped from the lower to the upper reservoir using the four machines as motors driving pumps. During the day the machines may be used for about four hours as turbines driving generators. The photograph shows the general site arrangement, including the upper and lower dams, connecting pipes and the power station.

(By courtesy of G.E.C. Machines Ltd.)

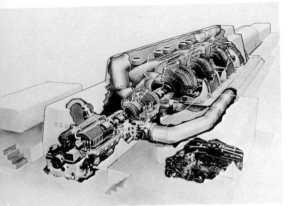

Cut-away view of 500 megawatt turbine.
(By courtesy of Reyrolle-Parsons Ltd.)

The turbo-generators at Blyth 'B' power station in Northumberland during commissioning. The metal casings in the foreground
enclose a complete turbo-generator set comprising a four-cylinder
steam turbine, 275 megawatt electrical generator and exciter
supplying direct current for the generator field. In the background a second set has the two low pressure cylinders of its
steam turbine exposed.

(By courtesy of C.E.G.B.)

compared with the very early nuclear stations which cost £170 per kilowatt —
the effects of the high inflation in recent years being ignored.

For a coal-fired station the running cost, i.e. fuel, maintenance, salaries, etc.,
is about 0.4p per kilowatt hour (kWh) whereas for the best nuclear stations
at present being built the figure is about 33 per cent of this. At first sight
it might appear that coal-fired stations are the 'best buy' on capital cost only,
but the cheaper running costs of the new A.G.R. nuclear stations may tip the
scales the other way. The ultimate choice depends on the sum of capital and
total running costs (including interest payable on the loan raised) for the
estimated life of the two types of station, based on the planned use of the
plant — 30 years for coal-fired and 25 years for nuclear powered. More
advanced designs of nuclear stations will become even more competitive —
note that the new Hartlepool nuclear station is near the Durham coalfield.

The published capital and running costs for power stations in the early
1970s are shown in Table 2.

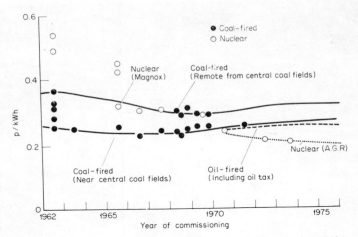

Running costs of nuclear and coal power stations (estimated 1970).
(By courtesy of U.K.A.E.A.)

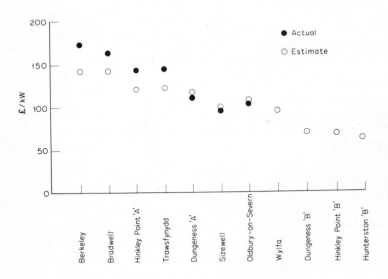

Construction costs of nuclear power stations (estimated 1970).
(By courtesy of U.K.A.E.A.)

Table 2

Power Station Costs (1972)

	Coal/Oil		Nuclear	
Capital cost	£40 − 55/kW		£85 − 170/kW*	
Running cost	West Burton	0.33	Wylfa	0.47
	Ratcliffe	0.33	Oldbury on Severn	0.48
	Didcot	0.37	Hartlepool (A.G.R.)	0.28
	Fiddlers Ferry	0.36	Hinkley Point B (A.G.R.)	0.26
	Fawley (Oil)	0.33	Heysham	0.31
(base load)	Pembroke (Oil)	0.32		
in p/kW	Kingsnorth (Oil)	0.33	*Exclusive of initial fuel charge	

These figures assume the following ground rules:
Interest 8%; 20 years life for Magnox, 25 years for A.G.R., 30 years for coal/oil fired;
Load factor 75%. Initial fuel charge included in nuclear running costs. U.K. heavy oil
tax included in oil station running costs.

The high cost for each unit (kWh) of electricity for Didcot and Fiddler's
Ferry is because they are remote from the coalfields and will be given a lower
proportion of the total load. In other words they will not be run as often as
the stations which produce electricity more cheaply.

Let us suppose the anticipated demand proved wrong — remember the
mistaken estimate made in the 1960s — and consider the two possibilities. If
the demand for electricity should prove greater than expected, what would be
the consequences? Since the demand would be greater than the supply there
would either have to be voltage reductions or certain areas would have to be
cut off, both of which would be very undesirable. If, on the other hand, the
demand were lower than the supply available, it would follow that capital
plant purchased was not being used. Unused capital plant is wasted money!
The unused plant would not be paying for itself and hence the price of
electricity would have to rise to bring in the necessary extra revenue. Some of
this however, could be offset by scrapping earlier the old plant on the system.
In any case, this varying demand must be met by plant of different efficiencies
and ages.

Over the last decade there has been a very large increase in capital investment in the fields of generation, transmission, and distribution. The situation
reflects not only the growing demand for electricity but the industry's decision
to purchase the plant which is cheapest to run and also to improve the security
of the supply by building up an adequate margin of plant. The increase in
capital investment is in line with the national objectives of economic expansion
envisaged by the National Economic Development Council.

In the final analysis, however, the decision on the type and siting of a power
station is usually a political one made by the currently elected government
after the full facts and choices are presented.

2 How Electricity is Generated

Electromagnetic Induction

All electric generators depend on the principle of electromagnetic induction discovered by Michael Faraday, which states that when *relative motion* exists between a conductor and a magnetic field, an electromotive force (e.m.f.) is induced in the conductor. For a conductor moving through a magnetic field, the magnitude of the e.m.f. depends upon:

(*a*) the strength of the magnetic field,
(*b*) the effective length of the conductor,
(*c*) the speed of the conductor through the field.

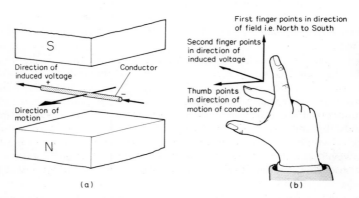

(a) Electromagnetic induction by the motion of a conductor in a magnetic field.

(b) Fleming's Right Hand Rule for the generator.

Once the conductor motion is stopped the e.m.f. falls to zero, but if the conductor is moved in the opposite direction the e.m.f. is reversed. Finally, the e.m.f. always reaches its maximum value when the conductor moves at right angles to the magnetic field. These are the fundamental facts of electromagnetic induction which apply to all electrical generators.

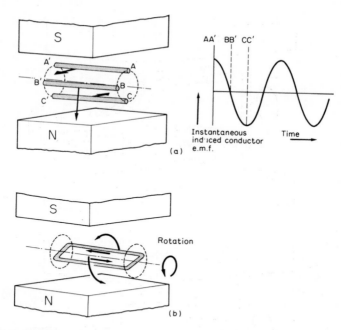

(a) Generation of an alternating e.m.f. by the rotation of a single conductor in a magnetic field.

(b) Single turn generator. The induced e.m.f.'s in the two long conductor sections are additive and cause an alternating current to circulate around the turn.

To generate electricity *continuously* one must move the conductor continuously through the field, for example by moving it in a circular path between the poles of a fixed magnet. As a result the induced e.m.f. acts first in one and then in the reverse direction, i.e. it pulsates or alternates.

A larger e.m.f. can be obtained by placing a second conductor diametrically opposite, connecting two adjacent ends together and the other two ends to sliprings. The resultant e.m.f. is twice that of the single conductor. Similarly, the use of many conductors connected in series in the form of a number of turns gives a proportionally higher output e.m.f.

The alternative to a rotating coil generator, and the one most commonly used, is the rotating magnet generator in which the coil is kept fixed and the magnet rotated. In the simplest possible arrangement a two-pole magnet spins inside a coil formed by two conductors, the e.m.f. measured across the fixed coil alternating as the magnet rotates. If a load were placed across the coil terminals, an alternating current would flow fluctuating in sympathy with the induced e.m.f.

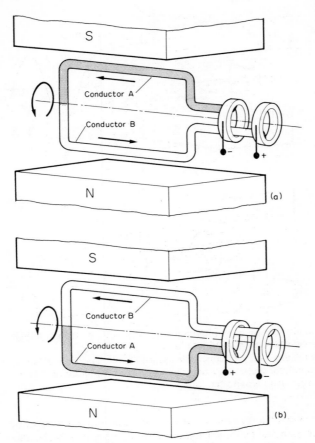

Single-turn alternating current generator with sliprings to collect
output. As the single turn rotates in the magnetic field the
polarities of the slipring collector terminals change sign.

Three-Phase Generation

Electricity is usually generated using a three-phase coil system. Let us look in
more detail at what this means and how it is obtained (see diagram, page 24).

Suppose three single-conductor generators are connected to a driving shaft
rotating at constant speed. As the north and south poles sweep across each of
the conductors, the induced e.m.f.'s in the conductors will be directed first
in one and then in the other direction, but the e.m.f. maxima will occur at

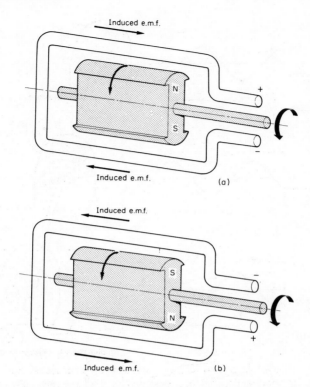

Rotating magnet generator. The alternating current output is
collected from the field coil inside which the magnet rotates.

different instants. The induced e.m.f. will reach a maximum in conductor AA′
in the diagram before that in conductor BB′ and in BB′ before that in CC′. A
rotation through an angle of 120° corresponds to 120 electrical degrees for a
two-pole generator, and it follows that the maximum e.m.f.'s in the three conductors follow each other every one-third of a cycle. For reasons of economy
of space and materials the three machines are incorporated into a single three-
conductor generator, with no change in behaviour.

The single conductors can, of course, be replaced by sets of conductors or
phases — designated by the colours red, yellow and blue — and the generator
becomes a three-phase generator. In practice all the conductors are mounted
on an iron core called the *stator* while six connections to the three phases
constitute the output terminals from which power is drawn.

A three-phase system may be likened to the three-cylinder internal combustion engine — similar to the type used in diesel-electric locomotives with the

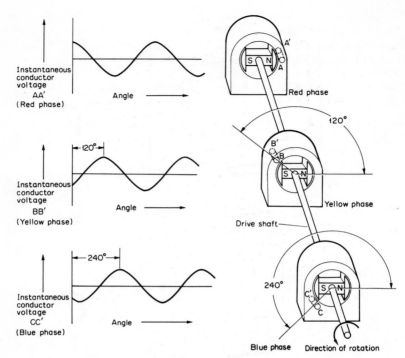

Three-phase generation using three separate single-conductor generators. Because the conductors are displaced physically by 120 degrees relative to each other, their induced e.m.f.'s are similarly displaced in electrical phase.

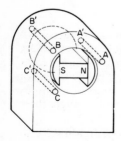

Three-phase generator with the three individual conductors placed in the same machine frame.

three cylinders (or phases) firing in the order red-yellow-blue. The engine gives a very much more uniform power flow to the driveshaft than a single cylinder engine, because each piston power stroke occurs at a different time during the cycle so that there is at least one piston undergoing part of a power stroke at any given time.

The three-phase method is now the standard method of generating electricity on a large scale. Three-phase generators in Great Britain and on the continent give output voltages which alternate at 50 hertz, whilst generators in North America and in parts of Japan give outputs at 60 hertz. A standard 500 megawatt British generator would usually have three-phase conductors producing currents of the order of 15 000 amperes at about 12 700 volts with a line voltage of about 22 000 volts.

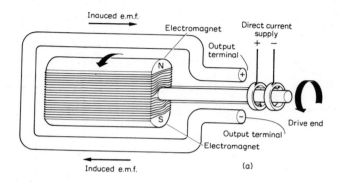

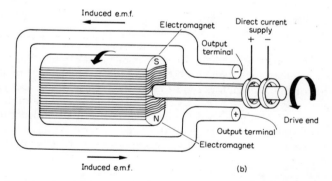

Rotating electromagnet alternating current generator. The rotor electromagnet is supplied with direct current via the sliprings.

The rotating member of a three-phase generator is called a *rotor*. Rather than a permanent magnet, the rotor is an electromagnet fed by direct current via a set of special brushes and sliprings. The rotating electromagnet construction does use a moving electrical contact, but since the rotor current is relatively small and of low voltage compared to the stator values, no great design problem is presented. Typical rotor currents and voltages would be 4000 amperes and 500 volts respectively, for a large generator.

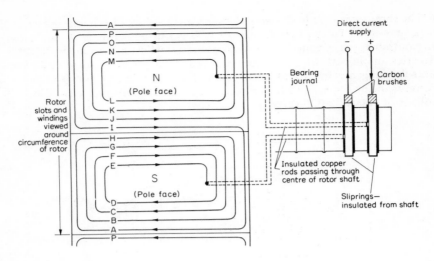

The electric circuit to give a two-pole rotor electromagnet.

Let us consider a few of the associated electrical and mechanical design difficulties with a view to explaining the advantages of a rotating magnet generator compared with a rotating coil generator.

As indicated above, high power electrical generators produce very high coil voltages, and it is extremely difficult to provide good electrical insulation for such coils if they are on a high-speed rotating shaft. The large centripetal forces acting on the conductors tend to squash the insulation, and vibration due to electromagnetic forces causes the insulation to wear by rubbing. The net effect is to reduce the thickness and quality of insulation leading to electrical breakdown. Moreover, with the large currents being delivered, there is severe sparking and burning of the sliprings and excessive wear of the collector brushes.

Neither difficulty arises with the rotating magnet generator and in addition it is easier to cool. Further, since the generating coils are fixed, they can be firmly braced into position, resulting in strong mechanical arrangements capable of withstanding the forces set up between the current-carrying conductors. Such forces will be especially large when the generator is delivering its full load current or when the output terminals are accidentally short circuited.

Generator Construction

A typical modern rotor takes the form of a uniform cylinder of specially forged high-tensile steel which is kept as small as possible to reduce centripetal forces. Even so the peripheral speed may be half the velocity of sound. There

A 660 megawatt turbine-line under erection at the manufacturer's works. The turbine comprises five cylinders: a single-flow high-pressure, a double-flow intermediate-pressure and three double-flow low-pressure cylinders. The high-pressure unit is shown on the right, followed by the intermediate and low-pressure units. Reheating of the steam takes place between the high and inter-mediate pressure units. All the shafts are solidly coupled to each other and to the driven generator. The particular set is the first of three to be supplied to Drax power station in Yorkshire.
(By courtesy of Reyrolle-Parsons Ltd.)

Rotor of a 500 megawatt generator before winding. Notice the deep slots and the relatively small overall diameter to reduce the centripetal forces on the rotating parts.
(By courtesy of Reyrolle-Parsons Ltd.)

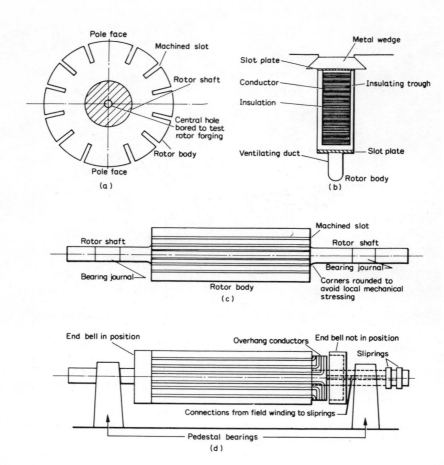

The rotor. (a) Cross-section through rotor shaft. (b) Conductors fitted in a rotor slot. (c) Side view of a rotor after machining. (d) Side view of a rotor after the electrical windings have been inserted. Notice the overhang conductors outside the rotor slots, forming the end connections. These must be retained in position by an 'end bell'. The rotor is shown without fans.

The very latest and biggest rotors have hollow conductors to enable direct hydrogen/water cooling, bigger pole faces, much longer rotor bodies and transverse/longitudinal slots in the pole faces to give the rotor equal stiffness along the two axes.

are a number of deep slots machined along the length of the body into which insulating troughs containing the field windings are inserted. The conductors are prevented from being thrown out at high speed by metal retaining wedges.

At the end of the rotor the conductors are held in place by snug-fitting retaining caps or end-bells, made of a special high-tensile non-magnetic steel to restrain the large internal bursting forces when the rotor is at top speed.

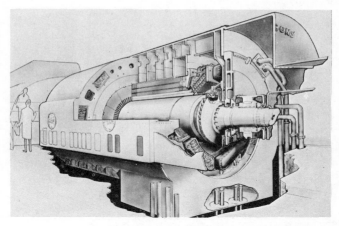

Cut away view of a 500 megawatt generator showing the
stator cooling circuits.
(By courtesy of Reyrolle-Parsons Ltd.)

The rotor field windings are connected to the rotor sliprings by insulated
copper rods passing through a central hole bored in the shaft. Bearings fed
with high pressure oil support the shaft and enable it to rotate smoothly.
When running at normal speed a thin film of oil is dragged around with the
bearing surface, which lifts the whole rotor a few hundredths of a millimetre
clear of the fixed bearing surfaces. In this condition there is no metal-to-metal
bearing contact and therefore negligible wear.

The stator is built up of a large number of thin steel sheets or laminations
firmly held together as the central core unit by two large end-plates. Slots in
the core receive the stator coils, which are thickly insulated in order to
'contain' the high voltage induced in them, and again these are held in position

Fitting the stator core into position in the stator frame of a 500 megawatt
generator. It is the heaviest lift in the turbo-generator set, weighing
200 000 kilogramme.
(By courtesy of Reyrolle Parsons Ltd.)

Fitting the wound rotor into the stator prior to works testing.
(By courtesy of Reyrolle-Parsons Ltd.)

by slot wedges. Ventilating ducts pass through the core so that the internally generated heat can be removed effectively by the cooling system. Large generators are cooled by hydrogen which, although a highly inflammable gas, is a most efficient medium for heat removal. Hydrogen is superior to air as a cooling medium for reasons discussed later. The stator coils are braced in position by substantial clamps where they leave the slots, and three-phase output connections from the stator coils are generally located underneath the generator. The core is supported by a fabricated mild steel frame which in turn is fixed to solid concrete foundations.

Principles of Generator Design

It is the generator designer's job to obtain the maximum possible electrical output from a machine of a given physical size using the materials that are available to him to the best possible advantage, and at the same time he must ensure safety and reliability. It would be useful at this point to settle the limits governing the generator output, assuming that the turbines can provide the necessary drive power. What, in fact, is the maximum power one can get out of an electrical generator without it breaking down?

To answer this, one must look at the factors affecting the electrical power output of a generator, namely the terminal voltage and the current produced. The limits must therefore be set by the maximum voltage generated and the maximum current that can be drawn satisfactorily, the former being limited by the effectiveness of coil insulation and the latter by the efficiency of the cooling methods.

It is convenient to regard generator design as pertaining to three main circuits — the electrical, magnetic, and ventilating circuits, and to consider each of these in turn.

The Electrical Circuits

The principles of three-phase generation have been explained and the electrical circuits of rotor and stator must now be discussed.

Rotor Circuit

The rotor of a turbo-generator is an electromagnet and therefore the rotor circuit is the electromagnet field winding. Since the poles of the electromagnet always have the same polarity direct current must be used, and this may be obtained in one of several ways. Perhaps the simplest is to have an entirely separate auxilliary power supply feeding the electromagnet by means of sliprings, but this is very expensive. An alternative is to employ an 'exciter' — a small alternating-current generator mechanically coupled to the main turbo-generator — feeding a static bank of semiconductor rectifiers via sliprings. The rectified direct current may then be fed to the rotor field coils using a further set of sliprings.

In the last few years an improvement in the latter technique has been introduced whereby the need for sliprings and associated brushgear has been removed. The 'brushless' excitation method uses an alternating current exciter, as before, but has its semiconductor rectifiers strapped to the high speed rotor. The rectified direct current is then fed directly to the rotor field coils. As for the rotor field coils themselves, they are of the 'concentric coil' type, which is a name given to them because of the way in which successive coils are fitted inside each other.

Stator Circuit

There are two aspects to be explained: (a) How groups of conductors are connected to form each of the three phases and (b) How the three phases are inter-connected to produce a suitable electrical output arrangement.

(a) Conductors can be arranged together to form phases in a number of ways, but their common feature is that the conductors are uniformly distributed around the core. The most common method of winding is to put an upper and a lower coil-side in each slot, so that the number of coils is equal to the number of slots.

The simpler arrangement shown on page 32 of a stator winding shows the inner cylindrical surface of the stator cut open and unwrapped to form a flat surface. The winding is for a two-pole eighteen-conductor generator with six conductors per phase. The 'starts' of the three phases are separated in the stator by an angle of 120° and on the diagram are indicated by the letters A, B and C respectively.

(b) The three-phase coils are usually connected by the method called the star-connection in the diagram of three-phase windings. The technique allows

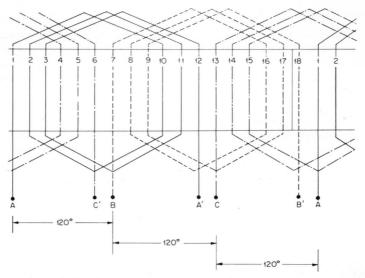

An exploded view of a stator three-phase winding. The three coils
have beginnings A, B and C, respectively and endings A', B' and
C', respectively. This type of winding has one coil-side per slot.
The coils are wound on diamond-shaped formers.

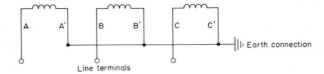

Three-phase 'star-connection' of phase coils.

two voltage ranges to be obtained from the generator. Terminal leads A', B',
and C' are connected to a common junction called the 'neutral', whose lead is
coloured green and connected to earth. This earth connection is necessary
because it facilitates the electrical protection of the generator, distribution
network and consumers' appliances plus the lightning protection of the transmission lines. Terminals A, B, and C are called *line terminals*.

The voltages measured between AA', BB' and CC' are referred to as 'phase
voltages' while the voltages measured between AB, BC, and CA are 'line
voltages'. At first sight it would appear that the line voltage measured across two
phases is twice that measured across a single phase, but this is not so because the
phases do not achieve their voltage maxima together. Line voltage is $\sqrt{3}$ or
about 1.7 times the phase voltage so that, if the voltage across each phase
were 12.7 kilovolt, the line voltage would be about 22 kilovolt, as mentioned
previously.

The Magnetic Circuit

It has been shown that a turbo-generator rotor acts as an electromagnet which produces the necessary magnetism for the operation of the generator. To be more specific, electric current flowing in the rotor coils produces a magnetizing force to drive magnetic flux around the magnetic circuit, hence linking the stator conductors.

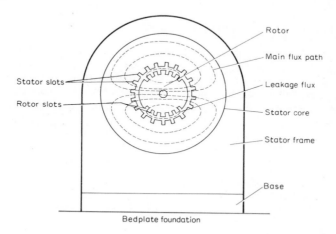

Cross-section of a generator showing the magnetic flux paths.

For a 3000 or 3600 revolutions per minute rotor, two magnetic poles are set up on diametrically opposite sides and when such a rotor is fitted into the stator, the resulting magnetic flux paths are as indicated in the diagram. The flux lines pass from the rotor across the air gap to the stator; most of the magnetic path lies in iron which can be highly magnetized. The air gap necessary for mechanical clearance between rotor and stator is kept small since it is difficult to force magnetic flux through air.

The term magnetic circuit or path has been used several times, and it is appropriate to qualify its meaning. This is conveniently done by comparing a magnetic circuit with an electric circuit.

Table 3

Comparison of Electric and Magnetic Circuits

Electric		Magnetic	
Electromotive Force	(E)	Magnetomotive Force	(F_m)
Electric Current	(I)	Magnetic Flux	(Φ)
Resistance	(R)	Reluctance	(R_m)
$I = E/R$		$\Phi = F_m/R_m$	

You are probably familiar with an electric circuit where an electromotive force E causes a current I to flow through a conductor of resistance R. The resistance of the conductor can be thought of as opposing the flow of current, the greater the opposition the smaller the current.

In a similar way, a current-carrying coil sets up a magnetomotive force F_m (where F_m = number of turns × current) which causes a magnetic flux Φ to be sent around the 'magnetic path'. It is this path over which the flux travels which is called the *magnetic circuit*. In its journey the magnetic flux is opposed by the magnetic reluctance R_m, which corresponds to resistance in the electrical case.

In theory, for any magnetic circuit the total magnetomotive force (m.m.f.) to produce the required flux can be calculated by finding the total reluctance of the circuit. If the circuit is made up of a number of parts through which the flux passes, the total reluctance is found by adding together the individual reluctances of the different parts. For a practical generator, the rotor field current produces the m.m.f. needed to drive the flux through the rotor body, across the air gap, down the stator teeth, and around the stator core to the other pole.

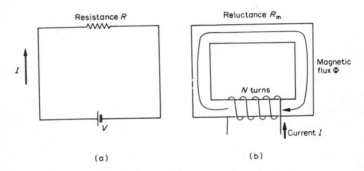

Comparision of electric and magnetic circuits.

You will notice however, that on reaching the stator the flux divides into two parts which travel along separate paths. (This corresponds to a simple, two-branch, parallel electric circuit in which the current divides equally into two parts.) The total reluctance of the equivalent magnetic circuit for the complete generator can then be calculated in a similar way to that used when dealing with series/parallel electric circuits.

In practice, the calculation of the reluctance of the individual parts is difficult because the reluctance of the iron parts depends on the operating flux density. Only the air gap has a constant reluctance because it is non-magnetic.

Instead of calculating the total reluctance of the complete magnetic circuit, the magnetizing ampere-turns giving the associated required flux density for

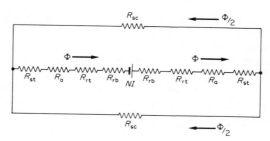

Magnetic circuit in a generator
NI = Magnetizing force in ampere-turns
R_{rb} = Rotor body reluctance
R_{rt} = Rotor teeth reluctance
R_a = Air gap reluctance
R_{st} = Stator teeth reluctance
R_{sc} = Stator core reluctance
Φ = Magnetic flux

each part can be added to give the total m.m.f. just as one may add up the individual volt drops around an electric circuit to give the total e.m.f. It turns out that about 95 per cent of the rotor ampere-turns are required for the gap even though it is small, the remaining 5 per cent being used for the rest of the iron path.

In the turbo-generator, and for that matter in magnetic materials employed in all electrical machinery, there are two main sources of inefficiency and unwanted heat losses, namely eddy current and hysteresis losses.

Eddy Current Loss

An alternating magnetic field sets up circulating electric currents in magnetic materials and to reduce these in the stator, it is made up of thin laminations. In practice, an insulating varnish is applied to both sides of each lamination before assembly, causing a splitting of the whole core into a large number of isolated sections. The net effect is to force the eddy currents to flow in high

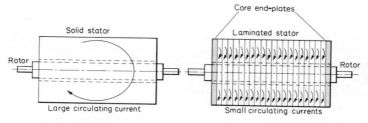

Side view of stator core showing induced eddy currents in solid and laminated types.

resistance paths, and hence they are considerably smaller than if the stator were solid metal.

Eddy currents also depend on the resistance of the magnetic material, so that the high resistance 3 per cent silicon steel permits only small currents to flow.

With such advantages one might ask why the stator frame and rotor are not made of laminations. In the case of the stator frame, the reason is because it does not or should not carry much magnetic flux. Mechanical strength is most important here. For the rotor the answer is not so obvious. Can you suggest why the rotor is not laminated but made of solid steel?

Hysteresis Loss

When magnetic material is taken backwards and forwards through a cycle of magnetization a loss is incurred called the hysteresis loss.

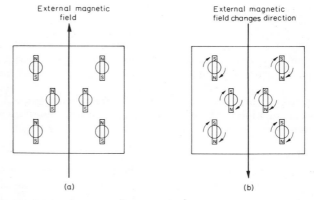

Hysteresis loss due to rotation of atomic magnets.
(a) Elementary atomic magnets align and their magnetism contributes to the main field.
(b) Elementary atomic magnets rotate to their new positions.

Atomic theories show that magnetic materials are made up of a very large number of atoms, each behaving as though it were a small permanent magnet. When an external magnetic field is applied across the material, these small magnets align themselves in the field, and their magnetism adds to that of the aligning field. If the external field is now reversed, the small magnets twist around into their new positions and in so doing experience a certain amount of 'internal drag'. If an alternating magnetic field is now applied across the material, the small magnets continually twist from the one alignment to the other, and energy must be continuously supplied to overcome the 'internal drag' acting on the small magnets, thereby producing heat.

The Ventilating Circuit

Because of winding-resistance, eddy-current and hysteresis heat losses, a generator becomes hot under normal working conditions. The heat losses due to winding resistance are most important because of the high currents employed. The current density in the stator phase coils can be as high as 1000 amperes per square centimetre. To prevent the machine from growing hotter and hotter, a ventilating system designed to keep the generator's temperature within the permitted limits of temperature rise must be employed. When the machine is working under normal operation at a steady load and temperature, the rate at which heat is generated internally is equal to the rate of heat extraction by the cooling system.

In small generators (up to 60 megawatt) air is used as a cooling medium, but on larger sizes hydrogen and/or water cooling is employed. One or two fans fitted to the rotor body circulate the cooling medium through the air gap and the rotor and stator ventilation ducts. The warmed gas is then directed past heat exchange coolers — essentially pipes containing cooling water — which remove heat from the gas.

The more efficient the cooling system, the greater the rate at which heat can be removed from the generator, and therefore the lower the temperature rise for a given load. It follows that a machine could be made to deliver more current, giving bigger winding losses, but with a temperature rise still within accepted levels, if it had a more efficient ventilating system, the only limitations then being the bracing of the electrical conductors mechanically against the large electromagnetic forces, and the supply of the necessary driving power. Bearing this in mind, let us consider the subject of ventilation in further detail.

Hydrogen Cooling

Why is hydrogen so effective at removing heat from a generator? The answers to this question lie in the properties of the gas and the physics of heat transfer. Three processes are involved:

(a) heat transfer from solid to gas,
(b) transfer of heat from one part of gas to another,
(c) storage of heat in gas,

Hydrogen is about 1½ times as efficient as air for the first process, 7 times as efficient for the second and 14 times as efficient for the third, so that the combined effect of these is to increase considerably the rate of heat transfer from stator and rotor surfaces to coolers. A considerable increase of machine output for the same temperature rise is therefore obtained by using hydrogen instead of air.

A further advantage of hydrogen relates to its effect on conductor insulation. Air-cooled machines, particularly those operating at high voltages, suffer deterioration of the insulation material due to corona discharge at highly

stressed points in the winding. (Corona discharge is the breakdown of air in
the immediate vicinity of electrified sharp metal parts.) In air corona produces ozone, and in the presence of moisture it produces nitric acid which
causes oxidation and corrosion. To a somewhat lesser extent, the insulation
which has been weakened already by the previous effect, also suffers from
the erosive effects of corona streamers.

Corona appears more readily in hydrogen because of its lower dielectric
strength, but this disadvantage is more than compensated by the complete
lack of oxygen and moisture in the cooling hydrogen. Hence the chemical
effects previously mentioned for air are inhibited. Deposits of dirt films,
bonded by oil and moisture, are also largely eliminated because of the use of
a gas-tight system.

Fire risks within the generator are also reduced, because a hydrogen
atmosphere of correctly regulated purity does not support combustion so that
damage caused by overheating, and consequent burning due to electrical or
other failure, is limited. Safety of operation is improved, and cost of repair
much reduced in the case of failure.

Nevertheless the use of hydrogen creates several challenging problems, the
major one being the risk of an internal explosion. Mixtures of hydrogen and
air are explosive over a wide range of proportions, from 5 to 70 per cent of
hydrogen by volume, but outside these limits they are non-explosive and do
not support combustion. For safe operation it is normal to maintain a high
state of purity in the generator, generally of the order of 95–97 per cent.

Operating facilities and techniques are so arranged that it is impossible for
an explosive mixture to form in the generator or in its auxiliaries. A special
purging technique has been developed which is applied when emptying and
filling the generator with hydrogen. First the air is displaced with carbon
dioxide, and then the hydrogen is admitted; when emptying the reverse
procedure is used. In case of accident the stator casing and doors are made
explosion-proof.

Direct Cooling

In traditional cooling methods the heat generated in the conductors had to
flow across a considerable thickness of electrical insulation before it could
escape or be removed by a coolant, and since good electrical insulating
materials are good thermal insulators too, only a small amount of heat
could flow through the insulation. Hence the current-carrying capacity of
the conductors had to be limited to prevent overheating.

During the last few years however, much success has been achieved using
direct methods of cooling conductors. In these methods, the gas is blown
along the *inside* of the current-carrying conductor so that the gas is brought
into intimate contact with the hot metal and the conductor heat is quickly
removed. Thus the conductors can carry much larger currents without
exceeding the specified maximum temperature rise.

An end view of the coils of a water-cooled stator winding. Each conductor is made up of a hollow rectangular copper strip through which cooling water circulates, and the conductor ends terminate in water boxes at each end of the stator. These water boxes are shown in the photograph with their covers removed, revealing the connecting links between top and bottom conductors.
(By courtesy of Reyrolle-Parsons Ltd.)

Water Cooling

Nowadays, it is becoming common practice to cool current-carrying conductors with water instead of hydrogen. As hydrogen has advantages over air as a cooling medium, so water has advantages over hydrogen. What do you think they are? Again, the water is fed down the inside of the current-carrying conductors.

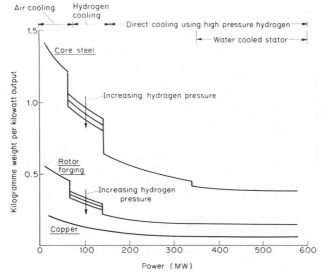

Reduction in material weights used in generators with increased set capacity.
(By courtesy of I.E.E.)

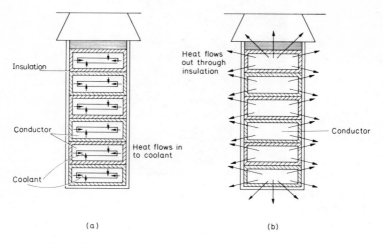

Heat flow patterns in (a) directly and (b) indirectly cooled
conductors (above).

Current-carrying capacity of air and hydrogen-cooled rotors
(below).

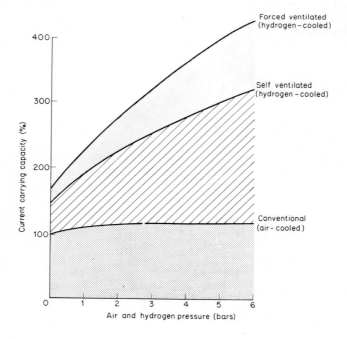

A typical system is shown in the diagram of a generator on page 29. Cool, demineralized water of low electrical conductivity is fed by large pumps to and from specially moulded insulated supports, called water boxes, located at each end of the stator frame. (Why should 'insulating' water have to be used?) These water boxes support the end winding of the conductor coils and allow water to be injected down the centre of the high voltage conductors. The heat extraction results in a temperature rise in the water of about 20 K at the far end of the conductor and since the temperature difference between the conductor and coolant is only about 5 K, the maximum temperature rise of the former is only 25 K. The low value of temperature gradient across the stator conductor insulation and the low rise of temperature of the conductor more or less eliminate problems arising from differences of thermal expansion in the stator slot.

Water cooling of rotors is already being tried, but there are many problems mainly concerned with passing water in and out of the rotor and the large water pressures built up because of its high speed.

Materials

Copper is used extensively in electrical circuits in generators, but aluminium is an alternative choice due to its relative cheapness and lightness. It has been used since 1949 in a number of rotor windings where its low density reduces the centripetal forces acting on the winding.

Carbon, chromium, molybdenum, vanadium and nickel are included in various proportions (up to 4 per cent) in the rotor steel to obtain the desired properties — the nickel improving the magnetic permeability and the others the mechanical strength.

High permeability for the stator core is achieved by using laminations made of cold-rolled silicon-steel. Eddy current and hysteresis loss can be reduced by increasing the proportion of silicon in the steel, but this results in increased brittleness and so a compromise must be made.

The electrical insulation of stator and rotor conductors must withstand permanently the electrical stresses imposed, often under severe conditions. It must withstand thermal gradients, sudden and sustained mechanical stresses, and chemical attack resulting from corona discharge or contamination by foreign bodies. Mica has been used as the main constituent of generator insulation in the past since it maintains its good electrical properties at very high temperatures, but it has the disadvantage that it is not very flexible except in thin layers. To overcome this disadvantage a form called Micanite consisting of thin mica flakes cemented together with shellac on to a backing medium has been used. This may be cut into strips and used as a tape or moulded into any required shape to fit into a particular slot. New insulating materials are constantly being developed, such as an asbestos-based insulation, and synthetic resin bonded-glass laminates.

Table 4

Comparison of Copper and Aluminium

	Copper	Aluminium (Electrical Grades)
Elastic modulus (N m^{-2} X 10^3)	11 000	6500
Melting point (0 °C)	1083	658
Mean specific heat between 1 °C and 100 °C (J kg^{-1} K^{-1})	394	922
Heat conductivity between 1 °C and 100 °C (J s^{-1} m^{-1} K^{-1})	382	230
Density (kg m^{-3} X 10^3)	8.9	2.7
Electrical conductivity at 20 °C (Ω^{-1} m^{-1} X 10^6)	56	35

Comparative Values (Copper = 1.00)

(i) Equal cross-section for aluminium conductor		
Weight	1.00	0.3
Electrical Conductivity	1.00	0.62
(ii) Equal electrical conductivity conductors		
Cross sectional area	1.00	1.6
Weight	1.00	0.485
(iii) Equal heating of conductors		
Cross sectional area	1.00	1.37
Weight	1.00	0.42
(iv) Metal cost for equivalent electrical conductivity cross sections		
Copper at £25/100 kg, aluminium at £20/100 kg.	1.00	0.37
Copper at £54/100 kg, aluminium at £20/100 kg.	1.00	0.17
Price of copper fluctuates widely: price of aluminium fairly constant.		

From these few examples it can be seen that materials used in this branch of electrical technology, as in other branches, are necessarily a compromise between electrical, mechanical, and other properties, including cost, and in a given case the various competing factors must always be weighed against one another.

See Appendix II.

Generator Synchronizing and Loading

The most important function of the central control room in a power station is to run up the generators to speed and then connect them to the National Grid. The process of connection is called *synchronizing*. In order that a generator may be synchronized, four prior requirements must be met: the generator frequency, terminal voltage, phase (the electrical angle) and phase sequence (the order in which the red, yellow and blue phases achieve their maximum voltages), must all be the same as on the incoming Grid. Synchronization is usually done manually with a special instrument called a 'Synchroscope'. If the generator is running too fast the pointer rotates clockwise; if it is running too slow it rotates anticlockwise. When the pointer is almost stationary, at the twelve o'clock position, the conditions are just right.

Control engineers usually prefer the synchroscope to be turning slowly in the 'fast' direction. The control switch which operates the three-phase circuit breaker connecting generator and Grid can then be turned smoothly to 'On', upon which the machine usually gives a slight quiver and synchronizing is completed. Since the machine is running 'fast' a small power output occurs immediately.

To make the generator deliver more electrical power to the Grid, extra steam must be fed to the turbine to drive it harder, but since the generator cannot go faster than the synchronous frequency of the Grid, the effect of driving the turbine harder is to advance the phase of the induced voltage of the generator relative to the incoming Grid voltage. The harder the turbine drives, the greater the advance in load angle and the greater the power output from the machine.

3 Transmission and Distribution

Grid Switching Compound

The power flow from the generator to the Grid lines is controlled by heavy-duty switches called circuit-breakers, devices capable of opening and closing a circuit in a fraction of a second. The title for this device is descriptive of its purpose, i.e. it can 'break' — or stop — a circuit quickly and safely.

Three-phase, 33 kilovolt oil circuit-breaker (below, left). The side view shows the tank lowered by a hand-operated winch. For proper operation the tank is filled with oil, immersing the contacts. The circuit-breaker is in the off position with the moving contacts fully separated from the fixed contacts. A set of double contacts is used for each of the three phases. Porcelain bushings insulate the incoming and outgoing high voltage conductors from the metal parts of the circuit-breaker and a central conductor through each bushing carries the current to the actual contacts.
(By courtesy of G.E.C. Switchgear Ltd.)

Three-phase 400 kilovolt air-blast circuit-breaker with pedestal-type current transformers placed on either side (below, right). The breaker has twelve interrupter units in series per phase which are permanently pressurized with compressed air. A compressed air receiver for each phase provides the base on which the support bushings are mounted. The breaker shown is one of a set in a C.E.G.B. grid switching station.
(By courtesy of G.E.C. Switchgear Ltd.)

When an electric current is switched off by separating two contacts on the normal domestic switch, an arc discharge is created between them, but since the supply voltage is only 240 volts, the arc is small and only lasts for a fraction of a second. At the very high voltages and currents used for transmission the power of the arc is considerable and must be quickly 'quenched' to prevent damage.

One type of circuit-breaker has its contacts immersed in insulating oil so that when the switch is opened by a powerful mechanical spring the arc is quickly cooled and extinguished by the oil (see Appendix III). Another type works by compressed air, opening or closing the circuit-breaker contacts and at the same time 'blowing out' the arc with a high pressure blast between the contacts (see Appendix IV).

Once the high current has been interrupted, two 'isolators' placed on either side of the circuit-breaker can be opened if required. Isolators are devices which cause a break in the circuit to isolate the breaker but which are themselves not able to interrupt high currents and voltages. The opening of the two isolators means that there is no chance of anyone receiving an electric shock whilst working on the circuit-breaker so that maintenance work can be carried out safely because the circuit-breaker is electrically 'isolated'.

From the circuit-breaker and its associated isolators the current is led to a set of busbars, or thick conductors, which run the whole length of the switching compound. Busbars acquired their name from two sources: the heavy connections were invariably *bars* (of copper or aluminium) and as they were used as gathering points for all the circuits, (for all = omnibus; from the

Night-time view of the large insulator testing station at Brighton (below, left) used by the staff of the Central Electricity Research Laboratory, Leatherhead. Here the behaviour of insulator assemblies under conditions of severe marine pollution is observed.
(By courtesy of Central Electricity Research Laboratories.)

Substation outside Ferrybridge 'B' power station in Yorkshire (below, right) showing the 275 kilovolt step-up transformer feeding power into the National Grid transmission lines via isolators.
(By courtesy of C.E.G.B.)

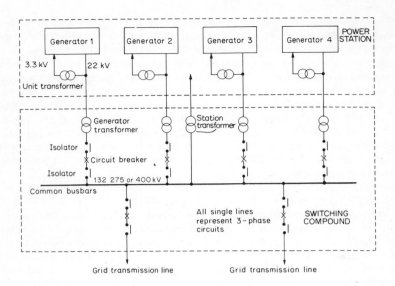

Electrical output arrangements from a typical power station.

The generator transformer supplies power from each generator to the common busbars in the grid switching compound. The circuit-breaker and two associated isolators in this line enable the power flow to the busbars to be controlled.

The unit transformer takes power from each generator to supply its own needs so that the set may be kept running if external supplies fail.

Two outgoing transmission lines with associated circuit-breakers and isolators, feed power into the National Grid.

The station transformer supplies auxiliary power to the station from the Grid when the station is shut down.

Latin) they could be regarded as 'omnibus bars' — anyway it's a good story and it may be true! Thus, although each generator in the power station has its own step-up transformer, circuit-breaker and isolators, all the electricity generated is fed on to the common busbars. From the busbars the power is directed to a second circuit-breaker, complete with its set of isolators, and out of the switching compound on to the Grid transmission lines.

Very High Voltage Transmission Lines

Transmission lines usually consist of overhead conductors supported from transmission towers by glass or porcelain electrical insulators. The three conductors from the three phases of the generator form a 'circuit'. Transmission towers generally carry two circuits, but to give some measure of protection against lightning an earthed wire is carried above the other conductors. At this stage let us ask why it is necessary to transmit electricity at a high rather

Standard 400 kilovolt tower carrying a double-circuit, three-phase
transmission line (right). Notice the 'quad' of conductors for each
line, the Stockbridge-vibration dampers and the overvoltage-
protection horns across the insulator assemblies. Lightning protection
is increased by the provision of a separate earthed line at the top of
the tower.

(By courtesy of B.I.C.C. Ltd.)

High voltage transmission lines (below). A pylon carrying a
400 kilovolt line can replace three pylons carrying 275 kilovolt lines or eighteen pylons carrying 132 kilovolt lines.

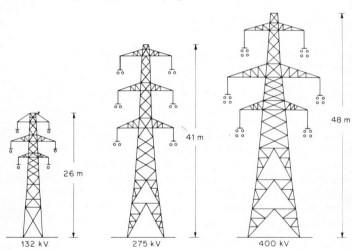

than a low voltage and, in fact, what determines the best voltage for transmission. The answers are based on economics. Let us first consider the two
types of cost, namely capital and running costs.

Transmission Line Capital Costs

(a) To transmit a given amount of power from generator to 'load centre'
requires a low voltage and a high current *or* a high voltage and a low current.

(b) Higher currents require larger conductors to carry them without overheating, and so larger supports are required.

(c) Higher voltages require increased clearances and hence higher and
heavier supports.

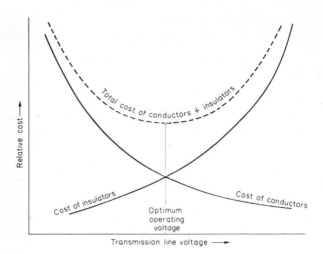

Transmission line capital costs. There exists an optimum operating voltage for which the installation cost is a minimum.

From points (a) and (b) it follows that if a high transmission voltage is selected, the current will be smaller, hence the necessary conductor will be smaller and therefore cheaper, but this is partly countered by point (c).

If the transmission voltage is too high, the extra cost of the very high-voltage supporting insulators and the difficulties entailed in providing extra clearances offsets the gain obtained by using smaller conductors. Modern Grid transmission lines operate at 400 kilovolt which is about the most economical voltage for long distance transmission in this country so far. Voltages up to 800 kilovolt are being studied to identify the problems to be met if the future growth of demand points to the need for such high voltages.

Table 5

Summary of Transmission Line Characteristics

Item	Type 1	Type 2	Type 3	Type 4	Type 5
Line Voltage (kilovolt)	132	132	275	275*	400
Number of Circuits/line	2	2	2	2	2
Number of Conductors/phase	1	1	2	2	4
Conductor size-equivalent area of copper (cm^2)	1.13	2.58	1.13	2.58	2.58
Normal power capacity per circuit (megawatt)	100	150	430	620**	1800
Standard span (m)	300	300	370	370	370

* This line is converted to 400 kilovolt by changing insulators.
** Capacity increases to 900 megawatt at 400 kilovolt.

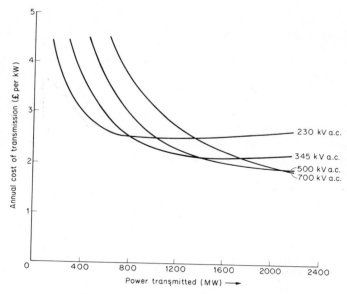

Transmission line running costs. For long transmission lines it is cheaper to use very high voltages.

Transmission Line Running Costs

There are two sources of loss which add significantly to the running costs:

(a) Heat Loss I^2R

When a current flows through a wire some power is lost in overcoming the resistance; this power is dissipated as heat. It should be noted that the power loss is proportional to the square of the current through a given conductor so that the current should be kept as small as possible for minimum heat loss.

(b) Corona Loss

At very high voltages the air surrounding conductors breaks down and begins to glow and crackle. This phenomenon is called *corona discharge*. The smaller the diameter of the conductor, the lower the voltage at which corona occurs, and hence the conductor's size is a vital factor. Corona discharge creates a power loss which, in very bad conditions such as damp, dirty atmospheres, can be quite significant, and it causes radio and television interference.

Corona loss is greatest when the voltage is high and the conductors have a small diameter. To reduce this loss, therefore, operating voltages should be made fairly low and the conductor diameter large. Naturally, there is an optimum voltage for which the effect on running costs of heat and corona losses taken together is minimal, but this voltage is usually higher than the optimum voltage based on capital costs. It is interesting to note that two small conductors spaced about 30 cm apart behave on the criterion of corona loss as a single conductor of diameter equal to this spacing. In other words such twin conductor systems have much lower corona losses than single conductor systems with the same cross-sectional area. This is the reason why 'twin' conductors are used on 275 and 400 kilovolt lines. On the very large capacity 400 kilovolt lines, four conductors per phase are used. These 'quad' lines, as they are known, have twice the current capacity of twin conductor phase lines with the further advantage that four conductors set in square formation have even lower corona losses than twin lines.

Types of Overhead Conductors

In selecting the type of conductor to be used for a transmission line one must consider not only the current to be carried but also the mechanical properties of the conductor material. The conductor must carry its current without overheating and must also possess high mechanical strength together with a low density, since it has to be suspended over long spans between towers.

The material with the highest electrical conductivity is silver but its use is prohibitive on the grounds of excessive cost. The next best metals are copper and aluminium. Copper is superior to aluminium as regards its current-carrying capacity but aluminium is both cheaper and lighter. Further, copper can be made quite strong using special manufacturing techniques, whereas aluminium is a soft ductile metal of low strength — about the same as good quality timber.

Fortunately it has been found that by combining steel wires with aluminium wires it is possible to make a composite conductor of the necessary strength which is not too heavy. The composite aluminium-steel cable is much lighter

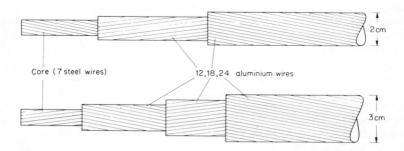

Core (7 steel wires)   12,18,24 aluminium wires

Types of overhead conductor. The diagram shows two standard, steel-cored, aluminium conductors in common use. The upper conductor has a breaking load of 78 000 newtons and the lower one of 130 000 newtons.

(By courtesy of C.E.G.B.)

than the equivalent copper conductor and therefore requires fewer supporting towers. In the case of 132 kilovolt lines, four support towers per kilometre are needed for aluminium/steel wires whereas six per kilometre would be needed for copper.

Statutory Requirements

In stringing overhead transmission conductors between towers, certain legal

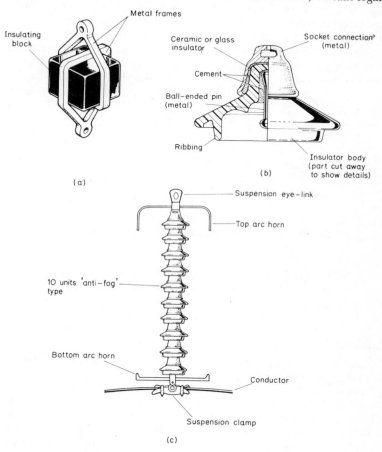

(a) An elementary insulator.
(b) A modern 'anti-fog' insulator unit.
(c) A typical 132 kV suspension insulator set. A 'set' comprises a 'string' of units as in (b), fitted together with their top and bottom fittings.

(By courtesy of C.E.G.B.)

requirements must be met with regard to the clearance distance between live electrical parts and earth, i.e. between the very high voltage conductors and the buildings, ironwork, trees and ground. This estimation can be difficult. In summer the overhead conductors expand and sag because of the heat, whilst in winter they become taut because of high winds, cold and ice formation. In the depths of winter when all the various factors are operating together the tension in the wire is particularly severe and safety factors must be adopted to prevent fracture in the worst possible conditions. These are incorporated in various regulations for the guidance of designers and the safety of the general public. Even winds of normal force can cause dangerous vibrations to be set up on the lines between the support insulators, which could cause fatigue at the highly stressed support points. Special dampers must be fitted near the insulators to damp out any resonant vibrations that might build up.

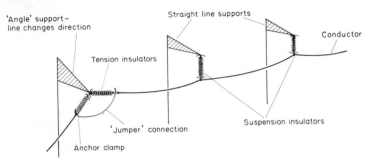

This diagram shows how insulator sets are used in service. Tension insulator sets facilitate the turning of a corner, the 'jumper' connection is needed to maintain the continuity of the conductor.
(By courtesy of C.E.G.B.)

National Grid

As indicated previously, the National Grid is not only a set of transmission lines from generators to load centres but a fully interconnected system. The British Grid, in fact, is the largest fully interconnected system under unified control in the world, and comprises about 180 power stations and 950 major grid switching stations interconnected by a network of very high voltage transmission lines. As a result of this interconnection, power stations which generate most cheaply can be used to the maximum whilst older and less economical stations are confined to peak demand periods, and so the country's generating capacity can be run at the most economical level.

Anglo-French Co-operation

The British Grid System is connected to two other electricity systems; it is connected to the South of Scotland Electricity Board's lines and to the

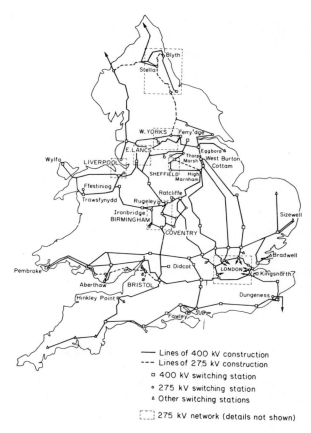

— Lines of 400 kV construction
--- Lines of 275 kV construction
▫ 400 kV switching station
○ 275 kV switching station
△ Other switching stations
⬚ 275 kV network (details not shown)

Main supergrid network 1970/71.
(By courtesy of C.E.G.B.)

Electricité de France network by a cross-Channel cable. The latter is a particularly good example of co-operation in load sharing.

The British quote prices per unit of electricity to France several times a day and they quote prices to us. The link runs from the cheaper to the dearer.

This link is interesting as it operates with direct current having transformers and rectifying valves at both ends. At the moment the line connecting the two countries permitting power exchanges operates at ±100 kilovolt to earth, and allows load sharing up to 160 MW. By controlling the voltage at the two ends of the cable it is possible to control the load flow to suit any conditions.

Are Transmission Lines Eyesores?

The problem of the Grid transmission lines is constantly appearing in the
news on television and in newspapers. The Generating and Area Boards are
sometimes accused of wilfully destroying the scenic beauty of rural landscape
by the erection of ugly transmission towers and ubiquitous lines. It is claimed
that the countryside is fast disappearing and what remains must be preserved
for amenity and scenic value. This case is a strong one and must be respected.
On the other hand, it can be argued that the Grid is essential if we are to
maintain and increase our standard of living and that the lines are vital to the
nation's economy.

Why not bury the lines as underground cables? This is not such an easy
solution as it first appears. Depending on voltage, a kilometre of cable may
cost from ten to fourteen times more than a kilometre of line — more about
this later.

Both Generating and Area Boards try to route their lines so as to cause as
little annoyance and unsightliness as possible, and employ a large number of
artistic advisors for this purpose. All possible advantage is taken of background,
of broken country and of screening afforded by hedgerow trees or woodland.
Where possible the route is taken through wooded valleys where the lines merge
into the landscape most effectively, and high skylines are avoided.

With regard to the transmission line towers themselves, the C.E.G.B. have
studied many arrangements but consider their standard lattice-steel tower to
be the most simple and effective. The matt finish of weathered galvanizing
also has the least conspicuous outline when seen against hillside or sky. Paint
of various colours has been tried from time to time but with no great success.
In the last few years a new low height 400 kilovolt tower has been developed
with a height only two-thirds of that of the standard 400 kilovolt tower.
Although it is perhaps visually inferior to normal towers, it may be usefully
employed to avoid cutting sky lines.

The Generating and Area Boards are also very concerned about the impact
of the terminations of these lines in grid and distribution substations on the
environment. Though a 132 kilovolt substation may only occupy between
$4000 - 8000$ m^2, a modern 400 kilovolt substation with provision for landscaping, can require 0.15 km^2 or more of land. Great care is taken over their
siting. Local authorities are consulted and advantage is taken of cover afforded
by natural contours or by woodland. As the voltages employed in substations
are very high, large clearances are required around the electrical conductors
to prevent flashovers. Because of this and of the large forces which may arise
in the event of a short-circuit, much of the equipment, circuit-breakers,
transformers, busbars etc. is inevitably massive. However, a recent redesign
of the largest type of 400 kilovolt substation has enabled the height to be
reduced considerably. Qualified landscape architects are employed to prepare
comprehensive landscaping schemes for these installations. They usually
include the construction of grass-covered embankments and the skilful planting of trees and shrubs so as to blend the site into the surroundings.

'Live'-Line Working

Recently new techniques have been devised enabling men to work on the high voltage transmission lines whilst they are still 'live'. The technique in the case of the very high voltage lines owes its success to a special *'Faraday-cage' suit* worn by the linesman; this has an exposed 6.5 mm square mesh of stainless steel wires 0.5 mm in diameter, woven into the cotton materials of his 'coverall' suit. It has zip and press-stud fastenings, with separate gloves and bootees, and completely encloses the wearer. From basic principles of electricity, no electric field can penetrate such a metalized enclosure and the worker is safe. A support boom, insulated ropes, and a bosun's chair enable the linesman to move from point to point on the line. Such an arrangement means that the lines do not have to be switched off for repairs. It is estimated that this may save at least £1000 per day for each line out of use and could, in certain cases, amount to £10 000 per day. The suit costs less than £100. The technique for the medium and high voltage primary distribution lines is to work from the ground wherever possible. Special tools are fitted to long insulated handles and support poles which enable the groundsman to perform his task in greater comfort and safety.

'Live-line' working at 400 kilovolt. A linesman in a 'conducting' suit prepares for ascent to live conductors to make a repair. (By courtesy of the C.E.G.B.)

Underground Cables

Many people ask why the transmission lines cannot be put underground.
The simple answer is one of cost. An underground cable is a complex and
expensive item and installation of a 400 kilovolt cable costs over £800 000
per kilometre. (Such a cable would have a diameter of 12 centimetres and
weigh about 36 kilogrammes per metre.) For an equivalent 400 kilovolt overhead line the cost would be only £60 000 per kilometre. The cost ratio is
smaller at lower voltages but overhead lines are still preferred.

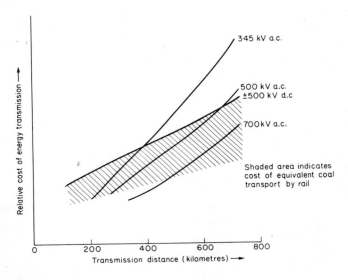

Costs of energy transmission with distance.

The cost of underground cables is high because of cooling and insulation
difficulties. When circuits are put underground, each conductor has to be
covered with thick electrical insulation built up from paper tapes which
prevents electrical leakage to earth or to other conductors. This has the
disadvantage of stopping the heat generated in the conductor from escaping
and therefore the heating process has to be reduced by employing much
larger conductors in underground cables than would be necessary in overhead
lines for the same current-carrying capacity. The underground circuit must
also be covered with a non-porous sheath of lead or aluminium to prevent
moisture entering the insulation and causing a breakdown. (In some cases
high-voltage cables are fed from special pumping stations with oil or gas under
pressure.)

Nor can such large and expensive cables be buried just below the ground
surface; special trenches are required with joint bays every 300 metres.

Usually twelve separate cables are needed to carry the same power as a double-circuit 400 kilovolt overhead line, and because of all the heat produced they cannot be laid together in a single duct. Four separate trenches are needed, each 1 metre wide and 1 metre deep, and about 4 metres apart. The total excavation would be at least 16 metres in width — as wide as a three lane motorway — which would cause enormous disturbance to the countryside whilst the work was in progress and add substantially to the cost of the installation. A recent major cable development does offer the possibility of some economies. In this case, six cables only would be needed to match the full capacity of a full duty 400 kilovolt overhead line, each cable being installed in a plastic pipe about 25 centimetres in diameter, through which cooling water is circulated. Pumping machinery and heat exchangers housed in substantially sized buildings are, however, required every 3 kilometres along the cable runs. Crops can be cultivated over the cable trenches but building and tree planting are not permissible.

There is also the problem of maintenance. Whereas damage on overhead lines can normally be traced and repaired within hours, repairs to an underground cable usually take much longer because faults are hard to locate and the work is often difficult. The cost of search and repair are therefore very much higher.

Nevertheless, the use of cables in large cities is often unavoidable; however, the lengths of run are made as short as possible. In certain cases where the impact of overhead lines on beauty spots would be severe, a financial sacrifice is often made and underground cables used. In the 1960s about 370 kilometres of line at voltages of 132 kilovolt and above were cabled purely to preserve the countryside.

Superconducting Cables

It has been suggested that underground cables made of superconducting metals or alloys would be able to carry extremely large currents without becoming hot because of the absence of resistance. The cross-sectional area of superconducting cables might be very small indeed, but at present such an arrangement is impracticable because of the high cost of cooling the conductors to the necessary low temperature.

Microwave Frequency Power 'Cables'

Some reputable scientists forecast that bulk electrical power transmission of the future will be achieved using a special type of hollow metal pipe, called a *waveguide*, transmitting very high frequency, very high energy radio waves. Frequencies would be as high as 10×10^9 hertz — about ten times the frequency of B.B.C. 2 television.

High Voltage Direct Current Transmission Lines

The cross-Channel cable between England and France works with direct rather

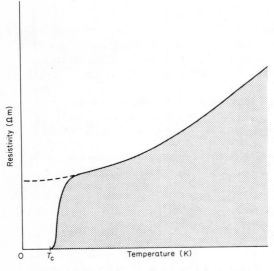

Electrical resistance of superconductors (above). The electrical resistance of superconductors suddenly disappears as they approach the absolute zero of temperature 0 K (−273°C). In some superconductors this may be in the order of 20 K but is usually much lower. This 'critical temperature' T_c is characteristic of the particular superconductor. The resistance of a non-superconducting metal, on the other hand, falls off smoothly to a minimum value (broken curve).

Cross-section of a proposed superconducting cable (below). (By courtesy of I.E.E.)

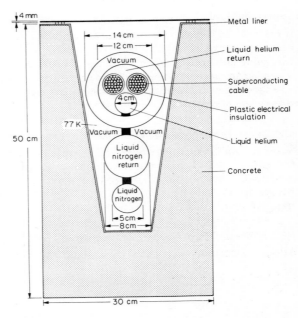

than alternating current. Perhaps you can explain why and suggest a few advantages of such an arrangement. High voltage direct current transmission has recently been the subject of much investigation and has been used, for example, to transmit hydro-electric power from New Zealand's South Island to the North Island via a 40 kilometre-long sea cable (+570 kilometre overhead line) and to link Sardinia to Italy by a 121 kilometre sea cable (+292 kilometre overhead line).

There are three main advantages of using such direct current transmission. An underground or submarine cable has the ability to store electric charge (called capacitance) in very large quantities and this naturally increases with cable length. A long cable therefore will have high wasteful 'charging currents' to an alternating current supply. These currents have to be supplied by the power stations feeding the system, particularly when the true load is low, and considerably increased costs result. With a direct current supply there is only the initial charging current at the switch-on.

Secondly, direct current enables the cable insulation to be more economically utilized than with alternating current. Electrical insulation must be able to withstand the maximum possible voltage on the line without breaking down which, on an alternating current supply, is $\sqrt{2}$ times the root mean square (r.m.s.) voltage. Any cable insulation must therefore to able to withstand not the r.m.s., but the maximum voltage, which is appreciably greater than the r.m.s. value. With direct current, however, the maximum voltage *is* the r.m.s. voltage.

Finally, direct current coupling of two systems of alternating current power obviates the need to synchronize them.

The disadvantages of direct current lines relate mainly to the necessity of a *converter* to convert from mains alternating current to direct current, and an *inverter* to change from direct current to the alternating current of the line. Converters and inverters are so expensive that direct current lines are justified only when the advantages to be gained are considerable.

Transformers

For medium lengths of overhead transmission line and short lengths of underground cable the alternating current line is to be preferred because the voltage used on a particular section of line can be set by means of a transformer to give the best economic advantage. Transformers are relatively inexpensive. You will be familiar with the transformer from previous studies. It is essentially two sets of coils wound on the same iron core. The input coil is called the primary; the output coil the secondary. The output voltage on the secondary side can be increased or decreased by increasing or decreasing the secondary turns. Windings and core are usually placed in a metal tank filled with oil for insulating and cooling purposes.

A three-phase transformer is basically three separate single-phase transformers wound on a specially formed iron core. You might at first think that

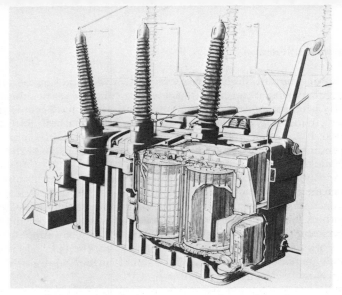

Cutaway view of a large power transformer (above).
(By courtesy of Reyrolle-Parsons Ltd.)

A large 400/132 kilovolt power transformer of about 200 mega-watt capacity ready to be lowered into its oil-filled tank (below). It is a three-phase transformer having five limbs and the view shows the 400 kilovolt windings with their terminals. The three 132 kilovolt windings are wound on the inside of the three 400 kilovolt wind-ings, with the terminals taken out on the opposite side. The use of five instead of the three basic limbs means that the top and bottom horizontal core sections can be halved in size, thereby lowering the overall transformer height. The metal frame holds the laminated iron core together. Notice the extensive use of paper as insulation.
(By courtesy of G.E.C. Transformers (Hackbridge) Ltd.)

such a combined core would consist of three separate cores, each carrying a primary, secondary and common return core for the magnetic flux, but in practice the return core is not needed, for if the transformer is balanced, the phases carry equal but out-of-phase currents and the resultant field at any time is zero.

Many engineering problems still remain to be solved in transformer design, although the major ones of the electrical, magnetic and ventilating circuits appear to be understood. Indeed two of these problems illustrate the wide range of challenge presented to the engineer.

Large Grid transformers can weigh over 300 thousand kilogramme and may require unusual transport facilities. Roads, bridges and transport vehicles have maximum weight limitations and the inconveniences caused to other road users cannot be ignored. In view of the size of this type of transformer railways are, in most cases, impracticable. In many instances the major parts, such as oil and bushing insulators, are transported separately. Transformers (and equally heavy generator stators) are usually carried on special road vehicles having as many as forty-eight wheels and are hauled by two diesel tractors, one at the front and one in the rear. These giant 'road trains' move on officially authorized routes and are escorted by the police.

Air cushion vehicles, based on the hovercraft principle, are employed to carry the biggest power plant across bridges by enabling their weight to be more evenly distributed, thereby saving strengthening costs. Specially constructed roll-on/roll-off ships have also been built to enable large indivisible loads to be shipped from one part of the country to another.

Another problem of transformers is that of noise. The noise or hum from a big transformer in a residential area can annoy many people. It is a form of 'pollution of the atmosphere' and as such must be reduced or eliminated by the transformer engineer, who must trace its origins and supply a remedy. It has been found that most of the noise comes from the phenomenon of 'magnetostriction'. When a piece of the iron core is magnetized along its length, the length increases slightly; this repeated magnetization occurs twice each cycle and for a 50 hertz transformer will give an associated 100 hertz hum. The periodic vibrations of all the iron laminations in the core are transmitted through the oil and base to the walls of the transformer tank.

Large transformers will therefore act as large sources of noise. Since it is extremely difficult to apply sufficient mechanical pressure to clamp the laminations and stop them vibrating, other methods of preventing high noise levels must be tried. Antivibration mounts may be used to support the core and the tank walls may be stiffened to give added resistance to vibration. Associated transformer fittings, such as control cabinets mounted on the sides of the transformer, must be designed to reduce any self-resonant effects. If all this fails there may be no solution but to enclose the transformer in a special acoustic building. This is a satisfactory technical solution to the problem, in that it leads to a reduction in noise level of 30—35 dB, but it is very expensive. Some more fundamental solution must be found. Do you have any ideas?

High Voltage Distribution Lines

Power transmitted to the load area at 400 or 275 kilovolt, is stepped down
by a large transformer at the Bulk Supply Point to a value suitable for distribution *within* the load area. Usually the voltage is reduced to 132, 66, 33 or
11 kilovolt in different stages, and from here onwards the Area Boards are
responsible for the distribution. They must ensure that each consumer has a
reliable supply and that the voltage at his terminals does not fluctuate by
more than ± 6 per cent of the declared voltage.

Why do you think the voltage stipulation is important? If the voltage were
allowed to vary unrestrained, how would this affect the consumer?

Overhead wires are mainly used for distribution lines at 132, 66, 33, 22, and 11
kilovolt, but where the area is heavily built-up the power is fed along underground cables.

There is no fixed pattern for primary distribution, the arrangement of
substations and transmission lines being developed as a result of the requirements of the area. For example, in London, where power stations are already
close to consumers, electricity is fed directly from the generating stations
into the distribution network at 66 kilovolt or 33 kilovolt without any intervening transmission lines. Connections are made to the National Grid,
however, just in case of breakdown.

Medium and Low Voltage Distribution Lines

The high voltage distribution feeder cables lead to local substations where the
voltage is reduced by other transformers still further to 415/240 volt. It is a
three-phase, four-wire system.

Three-phase, four-wire armoured cables lead from the local substation to the
consumers. Cables are laid in the ground about 40 cm to 80 cm deep and
covered with a layer of bricks for mechanical protection. At a point near the
consumer's premises a joint is made in the cable and a two-wire service cable
led off. The connection is made between one phase or line and the neutral,
which is earthed. Consecutive houses along the street are taken off each line
in turn and the common neutral. The voltage between each line and neutral,
or the phase voltage, is 240 volts.

Such a phase arrangement is used because this tends to balance the current
equally in each of the three-phase lines. If the phases carry exactly equal

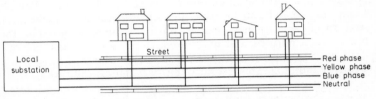

Three-phase, four-wire local distribution of electricity.

Three-phase 11 kilovolt cable. The conductors are insulated by layers of paper tapes tightly wrapped over each other and impregnated with oil. An overall sheath of lead prevents the entry of moisture. A steel wire mesh provides an 'armour' giving mechanical protection.

(By courtesy of B.I.C.C. Ltd.)

currents then there is no current in the neutral conductor, and the neutral wire could therefore be made very small. In practice, however, this would require an extra production line in the cable manufacturing process, and while at one time the neutral wire was half the cross-sectional area of the three-phase conductors, it was found more economical to make the neutral in a four-core cable the same size as the others.

Upon entering the house, the two-wire service cable is taken to a wall-mounted wooden board fitted with the Area Board's fuses, the Kilowatt-Hour energy meter, and the consumer's fuse-switch.

If a consumer requires a three-phase supply, so as to run a three-phase induction motor, for example, then a three-phase, four-wire service cable is tapped off the main distribution supply. The voltage between any two lines, or the line voltage, is 415 volt.

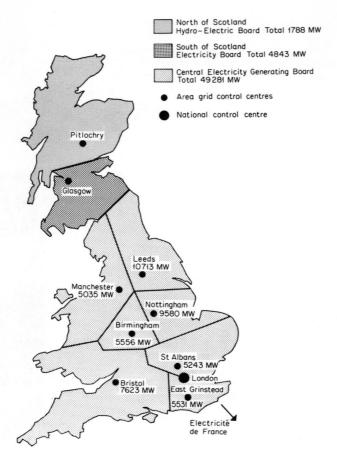

North of Scotland
Hydro-Electric Board Total 1788 MW

South of Scotland
Electricity Board Total 4843 MW

Central Electricity Generating Board
Total 49281 MW

● Area grid control centres

⬤ National control centre

Total power station output in each of the Grid Control Areas and
Scotland, 1971.

(By courtesy of C.E.G.B.)

Power Under Control

For the purposes of system control, England and Wales are divided into seven
Grid Control Areas, each with a Grid Control Centre, whose overall operation is
co-ordinated and supervised by the National Control Centre in London.

The seven Grid Control Centres are located in Manchester, Leeds, Nottingham,
Birmingham, St. Albans, East Grinstead and Bristol, and are each in direct
communication with all the power stations and grid switching stations within
their Area.

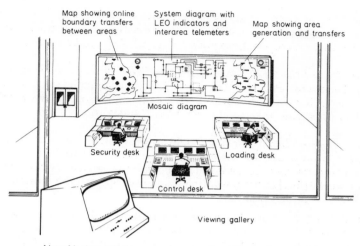

New National Grid control room. The operator is presented
only with the data he requires via cathode ray tube displays.
(By courtesy of C.E.G.B.)

At National Control there is a continuous watch over the electricity generation and flow patterns. National control has two major responsibilities:

(*a*) To ensure that the right amount of power is produced at the right time
in the most economical manner;

(*b*) To give maximum security of supply by ensuring that the National Grid
System has the greatest possible ability to withstand all probable failure
conditions.

Let us look at each in turn.

(*a*) *Power Production*

Power stations generate electricity most economically when they operate on
full load twenty-four hours a day. However, as the demand for electricity is
never constant, changes in consumption have to be balanced against changes
of production, so that starting up and closing down is necessary for some
generators during this period. Obviously generators vary in size, age and
efficiency. The newest and most efficient are the cheapest to run and should
be used as much as possible. The criteria for use of a given generator — the
so-called 'order of merit' will be discussed later.

National Control engineers estimate the national demand for electricity a
day ahead and arrange that the necessary generators are ready to take the
load. How are these estimates made and upon what does the demand
depend. From experience five main factors have been isolated.

(i) *The Time of Year*　In winter when there is less daylight and the temperature is lower, more power is used — mainly for lighting and heating, whereas at Bank Holidays there are sharp power drops as shops and factories are shut.

(ii) *The Day of the Week*　It is well-known that more power is required on Monday than on other days of the week. This is because factories, need more power to get going when they have been standing idle over the weekend, and housewives usually do their weekly wash on Monday too.

(iii) *The Time of Day*　There are wide power demand variations during a given twenty-four hours, and also the peak for the day shifts from evening in the winter to morning in the summer. The winter peak is, of course, due to the fact that shop and factory working hours overlap with the domestic load which comes on as darkness falls.

(iv) *The Weather*　Apart from being the leading topic of conversation in Britain, the weather is also a major factor in influencing the demand for power. Factors which increase the demand are a fall in temperature, increasing cloud cover, heavy rain, strong winds and fog.

(v) *Television*　Television is relatively new to power engineers and there is no doubt that it is a significant social force and an important factor in determining load. As may be imagined, its main effect is in the evening or at weekends when most people are at home. When there is a particularly popular television programme, its effect on demand is felt not at the beginning of the programme when the sets are switched on, but rather at the end of the programme; then a large number of people start to cook supper or boil a kettle and there is a sharp increase of demand in a short time. National Control engineers need to read the copies of the Radio and T.V. Times thoroughly every week!

Once the demand for the following day has been estimated, the machines which must be run to provide the necessary load together with some standby plant, can be decided upon. The cost of generating electricity is the main factor governing the choice of machine to be used. The decision is based on the 'order of merit' of each generator, which takes into account the efficiency of each steam turbine and the fuel costs for each particular power station. The individual generator sets are arranged in cost order with the cheapest first, and this is the order in which generators are started up on a rising load and shut down on a falling load. Once an 'order of merit' is decided on, it remains fixed until fuel prices change or new plant is commissioned.

The arrangement of daily plant ordering and of calculating the exchange of electricity between Areas is so complicated that the task is done using computers. A basic programme stores all the fixed data relevant to each generating set together with a numerical model of the Grid system. An estimate is fed in on the next day's load curve and the computer works out the best order for the day's plant.

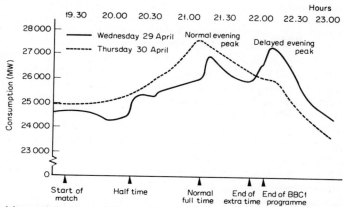

Television and the demand for electricity. Electricity for public supply cannot be stored in bulk, it must be generated at the instant of demand.

Television programmes, such as the Royal Command Variety Performance, the 'Miss World' contest, election coverage and some sports programmes, have a profound effect on public demands for power.

The big 'switch-on' comes at the end of television coverage as viewers all over the country put on fires, kettles, extra lights and other electrical appliances.

One example of the load distortions imposed on the system is illustrated graphically above. This shows the fluctuating demand during the F.A. Cup Final replay on Wednesday 29 April 1970, contrasted with the load curve for the following night (dotted).

(By courtesy of C.E.G.B.)

(b) Grid Security

The second responsibility of the National Control is to keep watch over the security of supply from the National Grid.

The Grid network in this country is the most extensive in the world. It offers many paths for power flow but some lines must be taken out of service for maintenance and the engineer must be sure when he does so that other lines are not overloaded. An engineer could work out these power flows with logarithm tables or a slide rule, but this would take several days. The answer is required in a few minutes in a changing situation.

The mathematical model of the Grid system expresses the layout and capabilities of the physical network as a series of formulae. Mathematical facts can be stored in the computer's 'memory' on magnetic tape. Engineers work out the day's load data and list any lines to be taken out of service and this information is punched on to 'paper tape'. The tape is then fed into a data transmitter which sends its information to the computer, and in a few minutes the answer comes back and is printed out. This answer shows the effect on parts of the system if further lines suffer a fault such as lightning. The computed security assessment is a valuable tool to National Control engineers.

National Control Centre

The Centre contains three main operational rooms—the National Control room, the Setting-in room and Teleprinter room.

In the minute-to-minute operation of the grid system, continual surveillance is necessary of system configuration, power flows, and voltage levels to ensure that the system will be reliable under all credible fault conditions. National Control require a substantial quantity of on-line information including more than 1000 telemetered readings and 7000 automatic indications of switch positions and alarms. With such a large amount of information to be marshalled and displayed, the assistance of computers has been employed, enabling this information to be stored and displayed selectively on cathode ray tubes as the need arises. In addition, traditional wall displays have been provided which summarize national minute-to-minute operating conditions.

The National Control engineers work at three specially designed control desks, each of which is provided with four cathode ray tube displays. (see page 65). Two of the desk are used respectively for system load dispatching and system switching duties, and the third is occupied by the National Control engineer who exercises overall supervision and control.

At each desk the control keyboards comprise an alarm and interrogation panel, and a tracker ball for rolling pictures depicting loading conditions on the grid system as required. The appearance is thus of a large map of the grid which may be moved in any direction at will, behind a fixed viewing aperture of limited size. Any one of the four cathode ray tubes per desk can be selected to bring up any computer display, and thus each control engineer has directly available four different simultaneous displays should he so require. Twelve simultaneous displays are thus available in the National Control room as a whole. The visual alarms comprise six red indicator lamps which flash when grid system alarms are received. The alarms are annunciated whenever grid switches operate, lines become overloaded, or when grid voltages or system frequency exceed prescribed limits. In acknowledging a given grid system alarm by pressing the appropriate alarm button, details of the cause of the alarm are displayed.

Additional facilities have been provided for the logging and recall of past data on the cathode ray tube displays. Samples of data are logged automatically in the event of grid system faults, periodically every five minutes and 'on demand'.

The on-line information received at National Control is updated continuously every thirty seconds. Certain complicated sections of the grid system associated with the main load centres require to be displayed separately and the control engineers are provided with facilities whereby they can call up a detailed representation of each of these sections. If an even closer look is required at individual switching stations within an area, these can be called up one by one to show circuit-breaker and isolator positions, and the power outputs of large generators.

In the *Setting-in room*, there are the input/output terminals to the computers which are used to assist the control engineers in the operation of the power system. Two types of computer calculation are required and these may roughly

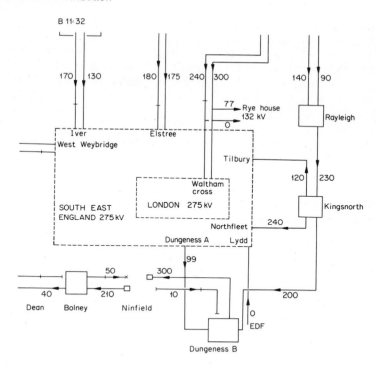

A computer-generated display showing part of the grid system.
(By courtesy of C.E.G.B.)

be classified as short-term predictive and immediate. In this context, short-term predictive calculations are those covering the period some few hours to twenty-four hours ahead; immediate are those carried out in real time concerned with minute-to-minute events as they occur on the grid system.

The *Teleprinter room* possesses teleprinter links between National Control, the seven Grid Control Centres and the Meteorological Office. There are also teleprinter connections to the South of Scotland Electricity Board and Electricité de France. The teleprinters can be 'ganged' together for communications which may need to be sent to several or all Grid Control Centres simultaneously. Weather maps are also received from the Meteorological Office in this room. The teleprinter system is used for interchanging with the Area Grid Control Centres details of estimated demands, economic power transfers, scheduled generating plant and transmission equipment outages, etc.

The on-line information display facilities described are provided by two process control type computer systems. These are connected in duplex configuration, one being the on-line master display computer and the other the standby computer, i.e. ready at all times to take over automatically the on-line display duties. However, with both on-line computer systems healthy, the standby computer will not run idle but will be used for the continuous on-line checking of grid system security by simulating the effect of credible grid system faults.

The on-line data required—telemetered flows, switch and overload indications, etc.—are sent to National Control via the seven Grid Control Centres over hired private Post Office lines.

4 Utilization of Electrical Power

If you have been fortunate enough to have looked around a modern factory you will have seen that electricity is put to work in four main ways, besides the obvious ones of providing light, heat and ventilation. Electricity provides:

Lighting in the new Mersey Tunnel (below, left). Over 3000 fluorescent lamps are used in the installation. The main in-tunnel lighting is provided by twin lamps, 2.5 metre, 85 watt fittings mounted above the two traffic lanes. Near the portals shown in the photograph, supplementary rows of single and triple lamp fittings are gradually introduced in separate sections. The number of lamps and fittings in each section is arranged to ensure quick adaptation of vision by drivers entering or leaving the tunnel. Lighting levels are automatically selected by a photo-electric system to suit the level of exterior daylight.

Outside the Liverpool-end portal the roadway is covered by an arch-shaped approach and a sun visor punctured with square apertures, providing a transitional section of combined natural and artificial light.

(By courtesy of Thorn Lighting Ltd.)

Electroplating car bodies (below, right). Total immersion lasts for exactly two minutes, and during this time a form of electroplating occurs. The negatively-charged paint particles seek out areas of positively-charged body with the least coverage and ensure a complete and uniform coating. Greatly improved protection against corrosion results.

(By courtesy of Ford Motor Co. Ltd.)

Two 2500 kilowatt direct current motors on test prior to despatch for installation in a hot strip mill at a new steel works in Yugoslavia. Their combined rating is 5000 kilowatt with speeds from 40 to 100 revolutions per minute. They will drive a reversing roughing mill capable of rolling 10 000–15 000 kilogramme slabs down to $2\frac{1}{2}$ centimetre thick.

(By courtesy of G.E.C. Machines Ltd.)

(1) *prime movers* in which electric motors are used to supply driving force.

(2) *process supply* where electricity is directly employed in one or a number of the manufacturing stages of a product.

(3) *control functions* where electricity is used to control method (1) or (2) either singly or in combination.

(4) *computers and automation* where the speed advantage of electricity enables it to perform complex tasks with the minimum of delay and without human intervention.

Let us consider these applications in turn.

1. Prime Movers

Electric motors as prime movers have many advantages over steam, petrol or diesel engines and they are highly efficient. This means that most of their input electrical energy is converted to output mechanical work. The small losses that do occur are due to electrical winding resistance, eddy currents and hysteresis of the iron core, and mechanical friction in the bearings.

Internal combustion engines, on the other hand, have large built-in inefficiencies. These cannot be eliminated just by avoiding friction and other heat losses, for the inefficiency is linked with the way in which the heat generated in the cylinder is converted to mechanical work. It can easily be shown that the efficiency is directly dependent on the engine's operating temperatures, and since these temperatures are fixed by factors such as the strength limitations of materials at high temperatures, the efficiency of this type of engine cannot be increased very much.

Electric motors have no such fundamental limitations. Further, compared with other motors they are essentially simple in design, robust in construction, compact in size and easily maintained. Being relatively cheap they are just about the 'best buy'.

The use of electric motors in factories is wide and varied. In the smaller sizes they are used for local dust extraction fans, and in medium sizes for lathes and other machine tools, while really powerful types are made to drive rolling mills in steelworks. In addition they provide the motive power for equipment used in the handling of materials and product movement on the assembly line, for example, lifting equipment, overhead runways, cranes, conveyors, elevators and even battery-powered industrial trucks.

Put yourself in the position of an engineer deciding on the sort of motor required to drive different machines and plant in a factory. How would you choose the motors you need? On what factors would you base your selection? Let us consider a few of the factors which might be relevant.

(a) Situation or Environment of Motor

If the motor is to be placed in the open air there will certainly need to be some protection to prevent the entry of moisture into the working parts. Motor ventilation arrangements should be such that they cannot be blocked by accummulations of ice and snow. Where motors work in very cold conditions they require special grease for their bearings. On the other hand, if the motor is to work in a hot environment such as in the roof of a boiler house it will require a better quality insulation to withstand the higher operating temperature. Motors which are to be used in hot and humid conditions in the tropics receive special treatment to deter the growth of fungus and prevent attack by termites.

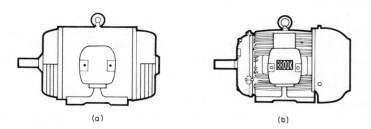

(a)   (b)

(a) Three-phase motor with drip-proof case.
(b) Three-phase motor with totally enclosing case for better
electrical and mechanical protection.
(By courtesy of Brook Motors Ltd.)

Suppose the motor is to be used indoors, will the ambient atmosphere contain dangerous gases or vapours? In potentially explosive atmospheres the complete motor will have to be totally enclosed, so that if a spark occurs in the windings the danger is isolated. Such is the case in chemical plants and coal mines. Total enclosure also prevents internal accummulations of dust and grit which constitute a further hazard — over a period of time the build-up of such debris would gradually affect efficiency and might eventually cause breakdown. Where motors are liable to occasional flooding in watery environments such as pump rooms, they must possess sealed bearings and the windings must be specially encapsulated; in extreme cases they may have to work completely immersed.

Besides the effect of environment on motors, there is also the effect of motors on environment. Although electric motors do not emit dangerous fumes they do emit a certain amount of vibration and noise. Such disturbances might not be objectionable in a factory machine shop but would not be tolerated for organ blowing in a church. Commutator motors, where frequent interruption of the electric current occurs, for example in an electric sewing machine or food mixer, cause electrical interference with radio, television and communication networks. When the interference is severe, disturbance to others can be avoided either by fitting suppressor units across the motor terminals which absorb most of the undesired electrical noise, or by the use of non-commutator motors.

(b) Mechanical Load

The *extent* of the mechanical load to be supplied by the motor determines its power, which is measured in watts or kilowatts. (The next few years will gradually see a decline in the use of the term 'horse power' applied to mechanical power in accordance with new recommendations. Mechanical power in these units was formerly converted to electrical power using the relationship 1 horse power equals 746 watt.)

The *nature* of the mechanical load may be continuous or intermittent. An intermittent load, where the breaks between working are long, does not cause the motor to get as hot as a continuously loaded motor, and in such cases it might be possible to use a smaller motor. On the other hand a motor heats up more quickly during starting, and if the breaks between working are short, overheating results. In such circumstances it is imperative to use a bigger motor than a continuous load would require. The overload capacity is determined by the maximum permitted temperature rise of the motor. A motor specification may state its power output over a short period, as well as the continuous rating.

The *characteristics* of the mechanical load are also important. Does the motor need to exert a strong turning effect or torque at high speeds or low speeds? It may even need to be strong at all speeds. Consider for example, the case of an electric motor suitable for a locomotive. A big effort must be exerted by the locomotive to start a train moving, but once the train is under way much less effort is needed because the force required at top speed need only be just sufficient to overcome the friction between wheels and track and the air resistance. Hence an electric motor for such an application must give a high torque at low speeds and a small torque at high speeds.

Full details of the mechanical load to be driven enables the planning engineer to fit the right motor to the task with the minimum in capital and running costs.

(c) *Power Supply*

The type of motor chosen must be based on the electricity supply available, i.e. whether a.c. or d.c. or a two- or three-wire system. Are the existing cable mains big enough to cope with the extra current demand? For large motor installations it might prove necessary to increase the cable capacity feeding the factory, in which case one might even consider bringing in a special cable to feed the motor or motors with a more economical voltage or system. Large motors may be designed to operate at quite high voltages in special cases such as those used for boiler feed pumps. Most motors, apart from the small sizes, require control gear to enable them to start smoothly and not cause too much disturbance to the mains voltage. The Area Boards have regulations on permitted motor starting currents, and financial penalties are involved if these rules are broken. Direct-on starting is usually only allowed for motors of 2.5 kilowatt or less on three-phase mains. The starting equipment, together with some means of overload protection, must be reckoned in with the overall cost of the installation. Direct-on starting of large motors, however, is permissible in power stations, because they are sited close to the generators, and in many factories and collieries which have heavy-duty electrical installations.

An important feature of most alternating current motors is that the current taken from the mains always lags behind the supply voltage. In technical language one speaks of a 'lagging power factor'. A big factory using many motors will, of course, take a large current having such a lagging power factor and electrical energy supplied in this way is more costly to the Area Boards than the same energy supplied where there is no lag. For this reason the Area Boards include a power factor penalty clause in the tariff. It is in the consumer's interest, therefore, to increase his lagging power factor towards unity. (Power factor is defined as $\cos \phi$ where ϕ is the electrical phase angle between applied voltage and resultant current.) One way of doing this is to install either banks of capacitors or a particular type of 'synchronous' induction motor which is dual-purpose, i.e. it can drive a machine *and* correct power factor. Many big firms save a great deal of money by installing equipment of this nature.

Now let us consider some of the types of motor available.

Three-Phase Induction Motor

The most important type of factory motor in use today is the three-phase alternating current induction motor. Such a motor consists of a stationary three-phase electrical winding mounted in a laminated iron core fixed in a support frame, i.e. the stator. The construction, although on a much smaller scale, resembles that of the stator of an alternating current generator. The electrical winding is supplied by three-phase mains.

The construction of the rotating part (or rotor), however, is different from that of an alternating current generator. The induction motor rotor has a slotted, laminated core into which solid, uninsulated copper bars are fitted. All the bars are connected together at their ends by short-circuiting rings and there is no external electrical connection to the rotor.

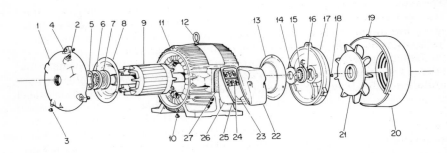

Exploded view of three-phase fan cooled induction motor.

1. Endshield, driving end.
2. Grease nipple and dust cap.
3. Grease relief screw.
4. Endshield securing bolt.
5. Preloading washer.
6. Bearing, driving end.

7. Rotating cap.

8. Flume securing rivet.

9. Rotor on shaft.
10. Drain plug.
11. Yoke with or without feet.
12. Eyebolt.
12. Eyebolt
13. Flume, driving end and non-driving end.
14. Fixed bearing cap.

15. Bearing, non-driving end.
16. 'Seeger' circlip.
17. Endshield, non-driving end.
18. Cap securing screw.
19. Lubricator extension.
20. Fan cover and three fixing screws.
21. Fan securing screw and fan key.
22. Terminal box cover and securing screws.
23. Conduit box gasket.
24. Terminal board and securing screws.
25. Conduit gland fixing screws.
26. Terminal box and securing screws.
27. Conduit bracket gasket.

(By courtesy of Brook Motors Ltd.)

When the power is switched on, the three-phase currents in the stator winding produce a resultant magnetic field which sweeps around inside the stator, causing induced currents to be set up in the rotor conductors. The action is similar to the way in which currents are induced in the secondary coil of a transformer when the flux linking the coil changes. Such rotor currents produce their own magnetic fields which react with the main stator sweeping field, creating a 'magnetic force' which turns the rotor.

The three-phase induction motor is very popular and has been called the 'work-horse of industry'. It is reliable in operation and cheap to buy but there are disadvantages associated with its use. The motor does not exert much effort at low speeds and speed control is difficult. Usually it is regarded as a fixed speed machine.

Over the years the size of induction motors has increased far beyond expectations. The situation is parallel to the growth of turbo-generators. Nowadays induction motors can have powers as high as 15 megawatt. In many cases induction motors are built as an integral part of the plant they drive, for instance, one advanced gas cooled reactor system uses blower motors actually sited in the high pressure carbon dioxide gas circuit.

A single-phase variety of the induction motor is used in the home for driving washing machines, extraction fans, refrigerators, etc. In this case the rotating stator field is achieved by using two separate windings fed from the

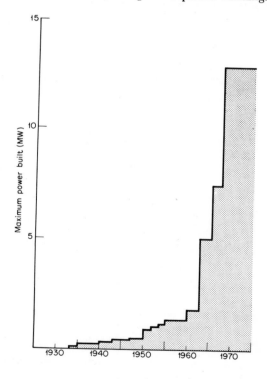

Growth in sizes of induction motors.

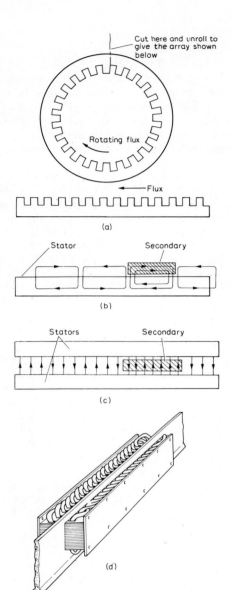

Linear induction motor
(a) Production of magnetic field with linear motion.
(b) Magnetic flux paths in a short rotor single-sided linear
    induction motor.
(c) Magnetic flux paths in a short rotor double-sided linear
    induction motor.
(d) Double-sided linear induction motor with a single sheet
    of conductor as the 'rotor'.
        (By courtesy of I.E.E.)

same single-phase mains supply but having a phase-shifted current in one of them.

In the last few years many ingenious types of electric motors have been invented for new applications. All the motors however, depend for their action on the same basic principles of electricity. Let us consider one or two new developments.

Linear Induction Motor

Most motors with which you are familiar cause motion which *rotates* a shaft. The linear induction motor is a type of three-phase induction motor which causes *linear* motion of the active element. Such a machine has been found useful, for example, for driving a shuttle across a loom. They are also used as accelerators by the U.K. Motor Industry Research Association at their car-crash testing station.

The principles of such motors were first tried in World War I when the Germans invented a tubular induction gun known as 'Long Tom', in which the shell formed a short rotor and the stationary stator coils were arranged so as to induce currents in the rotor and cause it to move with a high velocity out of the 'motor muzzle'.

During World War II aircraft launchers were built using a linear induction motor. The aeroplane was mounted on a trolley (acting as the short rotor) running on tracks along the stator block — about 300 m long. Later, because of the long length of track it was found more convenient to invert the roles of stator and rotor. The stator was made of a fixed copper ladder embedded in an an iron matrix, whilst the coils were made to move along tracks to carry the aircraft. During operation 100 MW of power was used. In a 100 m-long run, a jet plane was brought up to the take-off speed of 140 km/h in 4.2 seconds, and and the pilots using the device claimed that it gave a much smoother and more comfortable take-off than the conventional steam catapult. The new steam catapult however was better and is now used extensively.

Linear induction motors are now being planned to power high speed trains working at 400 km/h, the vehicle taking the form of a sort of tracked hover-craft. Experimental tracks have been set up in this country as well as in the United States, U.S.S.R., France and Japan.

Superconducting Motor

In 1970 the first really large superconducting motor was installed at the Fawley power station for an experimental period. The machine, designed to drive a water cooling pump, has a power of 2500 kilowatt and a running speed of 200 revolutions per minute and was built by the International Research and Development Co. — a subsidiary of Reyrolle-Parsons Ltd. A conventional motor for this application would need to run much faster and would therefore require an extra speed reduction gearbox to give the correct output. Such a combined motor and gearbox would be much heavier and bulkier than an equivalent superconducting machine.

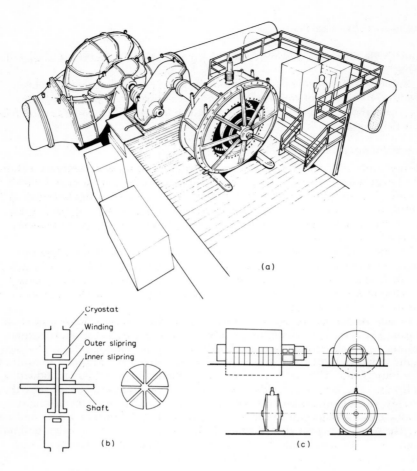

(a)

Cryostat
Winding
Outer slipring
Inner slipring

Shaft

(b)

(c)

Superconducting motor

(a) The drawing shows the 2500 kW superconducting motor which drives a main coolant-water pump at Fawley power station; the pump was designed to be driven by a conventional motor at 900 rev/min through a 4½ : 1 reduction gear box, but is now driven directly by the 200 rev/min motor.

(b) (left) Cross-section through a superconducting motor. (right) Copper homopolar rotor, showing segmented construction.

(c) Size comparison of conventional (above) and superconducting (below) motors giving 6000 kW at 50 rev/min; weight, cost (based on 1970 estimates) and efficiency comparisons are: *Conventional motor*; weight 370 000 kg, cost £220 000, efficiency 94 per cent.
Superconducting motor; weight 40 000 kg, cost £130 000, efficiency 97 per cent.

(By courtesy of I.E.E.)

The superconducting motor is of the homopolar type, using a 'normally' conducting 'Faraday-disc' rotating in a large magnetic field provided by a superconducting electromagnet. Direct current is fed to the disc by means of brushes sliding on axle and disc sliprings. The superconducting electromagnet coils are made from wire of a niobium-titanium-copper alloy cooled to liquid helium temperature (4.2 K) by a closed-circuit refrigerator. Because the superconducting wires have no resistance, very large direct currents can be passed through them without loss, and hence solenoids are obtained which can produce large magnetic fields. The Fawley motor has a magnetic flux density of about six times that of a conventional motor. Naturally, the main disadvantage of any superconducting motor is the need to provide refrigeration equipment, cryostats, and liquid helium etc., but even allowing for this extra weight, a superconducting motor of say 6000 kilowatt, would only be one-tenth the weight of a conventional machine.

Let us turn now to the second application of electricity in industry.

2. Processing Supply

The uses of electricity in factory processes are wide and varied and cover the cases where electrical energy is directly involved rather than being converted to mechanical energy in a motor. For clarity we will consider one particular process — electroplating — and a set of processes involved in a typical production line — the continuous casting of steel.

Electroplating

This was one of the earliest uses of electricity but in the modern world electroplating is carried out not only for decorative purposes but as a form of protection against corrosion. Typical examples are:

Automobile: Lamps, bumpers, hub caps, handles and lock parts, windscreen frames, radiator grilles.
Building: Door and window fittings, taps and sanitary fittings, hardware.
Domestic: Kitchen equipment and appliances, cutlery, furniture fittings.
General Engineering: Nuts, bolts, screws, piston rings, hydraulic fittings, electrical components.

Plating thicknesses are generally very thin — e.g. 0.06 millimetre — although heavier coatings of metals like chromium may be applied to give increased wear resistance. For a given material the electrodeposition is proportional to the plating time and current density.

The article must first be thoroughly cleaned before it is electrically connected to the negative side of a direct current supply and made the cathode. The anode, connected to the positive side of a direct current supply, is usually made of the metal being deposited, but in certain cases inert anodes may be used. The electrolyte may be an acid salt or cyanide solution. A low-voltage direct current supply at 4—12 volt is required for the process, and suitable plating currents range from 100 to 1500 amperes per square metre of surface area to be plated. This low voltage, highcurrent supply is usually obtained from banks of semi-conductor rectifier diodes fed from the alternating current mains.

A large amount of manually controlled electroplating is still done; one application is the selective electroplating of worn parts, such as bearing shafts, where the cost of stripping down to replace the part is high. However, most electroplating is now automatic, and apart from the saving in labour costs, the quality of the plating is consistently maintained. Losses through faulty processing and damage caused by handling are minimal. Automatic plant is economic for a throughput of 200 to 300 medium-sized articles per hour, and also when the articles to be processed have to pass through twenty or more tanks. Programme-controlled plants are employed in cases where a high degree of flexibility is needed.

One of the latest developments in electroplating is the production of dense metal meshes made by a special technique in which a piece of ordinary household plastic foam is used as an electrode. Deposited metal fills the interstices in the foam, which is then removed by burning, leaving a mesh of fine metal filaments. Such meshes may have applications in coal mines as a 'flame stop' where they could be used to absorb heat and prevent further ignition of an exploding gas travelling along a tunnel but not preventing gas expansion. The meshes could also be used to eliminate hydrocarbons from internal combustion engine exhausts, and they are an excellent soundproofing material.

Continuous Casting of Steel

The steel industry is concerned with making steel of the correct composition and right size in large amounts at a low price. To achieve this end, new methods of casting have been invented which reduce the number of manufacturing stages as well as the operators needed to keep the plant going. One of these methods is the continuous casting technique which attempts to make the finished steel product directly out of the molten steel. There are now well over a hundred big plants of this type in commission. The older and more conventional methods are slow, laborious and relatively expensive, for the hot ingot from the steelmaking furnace has to be rolled backwards and forwards many times in giant rolling mills to reduce it to the required size.

Let us look at the sequence of events in the continuous steel-casting process noting the multiplicity of ways in which electricity is used. From the store yards the scrap, pig iron and/or blast furnace hot metal are charged into the hearth of an electric arc furnace. Large electric currents are passed through the metal which becomes white hot and then melts. This is followed by the refining period when small quantities of iron oxide are added. A ladle conveys the molten steel to the steel-casting plant by means of an electrically-powered overhead crane. From the ladle the steel is poured into a big tundish which holds sufficient steel to even out any irregular flow, and a stream of molten

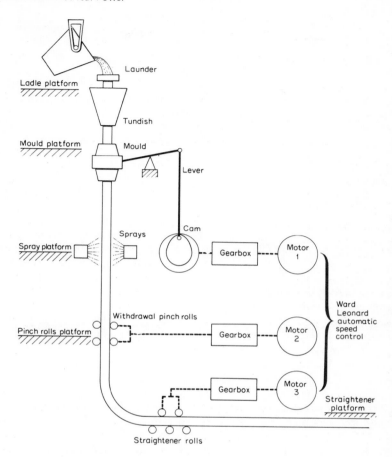

Continuous steel-casting machine.

metal spurts out of the bottom directly into the casting machine mould. This mould is water-cooled and in turn cools the outside of the molten metal so that it forms a strong skin to contain the remaining liquid. The mould is driven by a lever and cam in such a way that on the down stroke the hot metal in immediate contact with the mould cools and contracts, whilst on the up-stroke the contracted metal is free from the side of the mould and the mould can move unimpeded. Just below the mould jets of water are sprayed on to the hot liquid-filled billet so as to solidify it completely. The solid billet then passes between the withdrawal rolls which pull it from the mould. Finally, the billet is passed through straightening rolls and discharged horizontally for cutting into the required lengths.

Table 6

Simplification of Steel-making Process by the Continuous Casting Method

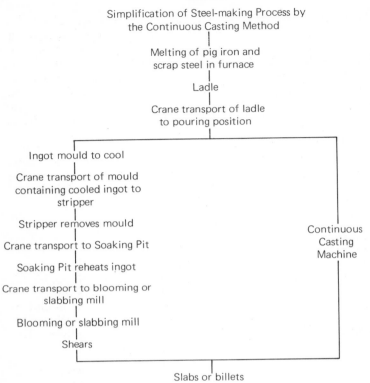

Melting of pig iron and
scrap steel in furnace
|
Ladle
|
Crane transport of ladle
to pouring position

Ingot mould to cool
|
Crane transport of mould
containing cooled ingot to
stripper
|
Stripper removes mould
|
Crane transport to Soaking Pit
|
Soaking Pit reheats ingot
|
Crane transport to blooming or
slabbing mill
|
Blooming or slabbing mill
|
Shears

Continuous
Casting
Machine

Slabs or billets

N.B. Slabs are of oblong cross-section greater than 8 cm X 4 cm; blooms are of square cross-section greater than 12 cm sides; billets are blooms that have been rolled to a smaller cross-section.

Electric arc furnace, capacity 110 000 kilogramme, in a Yorkshire steelworks. This furnace is part of the world's largest installation, comprising six units of this size and providing a total annual output of $1\frac{1}{2}$ thousand million ingot kilogramme of steel. The installation has a maximum demand of 150 megawatt of power supplied direct from the 275 kilovolt Grid.
(By courtesy of The Electricity Council.)

In everyday practice a continuous steel-casting machine is required to produce a whole range of product sizes, and each of these different sizes will require a different casting speed. Thus the motor driving the mould and the withdrawal and straightener rolls must all be adjustable, and once the arrangement has been set to a given speed, the speed must remain constant. This is difficult because the load varies rapidly during the cycle of operations. Such load changes are brought about, for instance, by the reciprocating action and changes in metal flow rate. Speed matching of the mould and the withdrawal and straightener rolls are also vitally important.

The Ward Leonard Motor Speed Control system satisfies all these requirements. It provides a system for adjusting the speed of the drive motors over a wide range and it gives the facility of automatic speed control under varying drive loads once the required speed is selected. This technique will be described to illustrate the third way in which electricity is put to use in industry, that is in its capacity as a control function.

3. Control Function

Servomechanisms

In the early decades of the eighteenth century the slow acting Newcomen beam steam engines needed an operator on continuous duty, alternately to open and close two valves operating the stroke of the piston. This monotonous manual task had to be done at the end of each forward and backward motion. The 'engine driver' acted as a sophistocated slave telling the engine what to do so that it could work in the correct way.

James Watt steam engine with flyball speed governor. From the output wheel a pulley drives the centrifugal pendulum linked to a throttle valve in the engine's steam supply line. When the engine speed increases the throttle valve reduces the steam to the engine; when the engine speed decreases the throttle valve increases the steam to the engine.

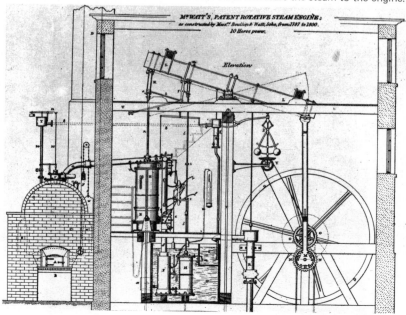

By 1719 technical advances eliminated the need for the man and an automatic control system was devised by modifying the Newcomen engine so that a mechanical coupling existed between cylinder valves and the swinging beam. The mechanical coupling 'signalled' the water and steam valves when to open and close. Here the coupling, not the man, acted as the slave. In 1787 James Watt adapted a governor to produce automatic speed control for his steam engines. A device which acts as a link to give automatic working of an apparatus is called a control system.

Examples of control systems are found not only in mechanical systems but also in purely electrical or combined electro-mechanical systems. In general control systems may belong to one of two forms, called the open- and closed-loop types. Open-loop systems are simply machines which carry out an order,

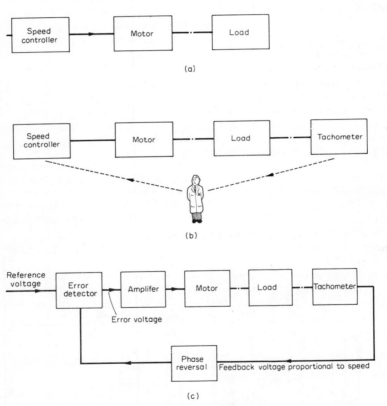

Control system for motor speed control
(a) Open loop
(b) Human closed loop
(c) Electrical closed loop

job, or function on command, without checking whether the order has been carried out. The sorting of articles on a conveyor belt into different feeder conveyors is a typical example; the sorting operation is performed on command from the photocells working via the power amplifiers on the mechanical handling plant, but once the operation is performed there is no check on the output to see if the execution is correct. On the other hand, a closed-loop control system makes a comparison between the actual output and the required output. If there is a difference between the two, the error is fed back to correct the input in such a way as to produce the required output. In practice the error signal is small, and additional amplifiers are added to ensure adequate power. It should be noted that it is theoretically possible to convert any open-loop control system into a closed-loop control system by the additional feed-back loop and its associated equipment. (See also Appendix V.)

A completely automatic closed-loop control system is called a servo-mechanism. Many diverse forms of servomechanism exist inside and outside industry today, although they might not be immediately recognised as such. Servomechanisms are used for such differing applications as the precision machining of metal parts from a master template to the automatic guidance of aircraft and missiles. Another application, already met with and to be discussed later in greater detail, is the Ward Leonard automatic motor speed control system, which is to be found in a large number of steel rolling mills. This system is an electro-mechanical equivalent of Watt's simple mechanical speed governor, i.e. it is a more sophisticated method of keeping an output shaft turning at a fixed speed independently of the load it is driving.

H. Ward Leonard first described his system in articles and papers presented as early as 1891 and 1892, but it was only in 1900 that he introduced his idea in a spectacular demonstration as a drive for a moving pavement. At the Paris Exposition he provided a moving 3 kilometre closed circuit of 2 parallel pavements driven at 7 and 14 kilometre/hour by 180 4-kilowatt motors. He showed that a fixed speed was maintained independently of the number of 'pedestrians' using the pavement and that the speeds of all the drive motors could be correctly coordinated.

Nowadays the Ward Leonard system is used in many industrial and other applications where such characteristics are required. For example, it may be used as a drive for colliery winders, travelling cranes, newspaper printing presses or gun turrets in warships as well as in steel rolling mills.

Control System Components

There are five main components involved in control systems. They are described below.

(a) *Monitor or transducer of the process output**. Any closed-loop control system has a sensing device which observes what is happening at the output

---

*See Appendix VI.

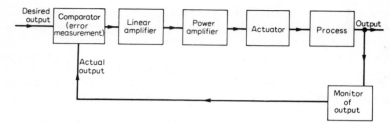

General servo-control system.

of the machine or process. This sensor or detector may be designed to measure speed, weight, temperature, dimensions and even colour! Sensors are sometimes called *transducers* because they *trans*form or in*duce* the relevant information into electrical signals whose strengths are proportional to the quantity being measured.

(*b*) *Comparator.* A control system comparator receives the electrical signal from the transducer, which gives a measure of the actual process output and compares it with the required output. If a difference exists between the two the comparator feeds a proportional error signal to the amplifier.

(*c*) *Amplifier*†. The amplifier receives the error signal from the comparator and raises it to a level sufficient to operate a power amplifier. For a linear amplifier the output voltage is directly proportional to the input voltage.

(*d*) *Power amplifier*†. A power amplifier then produces an output to operate directly the actuator which is placed to control the process input. (A power amplifier is required because the previous amplifier does not usually give sufficient power itself to drive the actuator.)

(*e*) *Output actuator to control the process input.* The powered actuator causes physical action to be taken at the process input to return the process output to its required value. Actuators may include such devices as electrical relays, motor-operated resistors or autotransformers, or motor-operated valves controlling the flow of water, oil, gas or steam. (Naturally the feedback to the actuator must be of the right polarity or phase so that the actuator motion *reduces* the deviation of the output from the required value.)

Fail-Safe

An important feature of all good control systems is that they incorporate the fail-safe principle. This means that should any malfunction occur, such as an electricity supply failure, then the equipment automatically renders itself safe by shutting down. A good example is the flame failure device fitted on central heating oil boilers. The circuit and relay are designed to allow a mechanical spring to shut off the oil valve automatically if a mains power cut occurs.

†See Appendix VII.

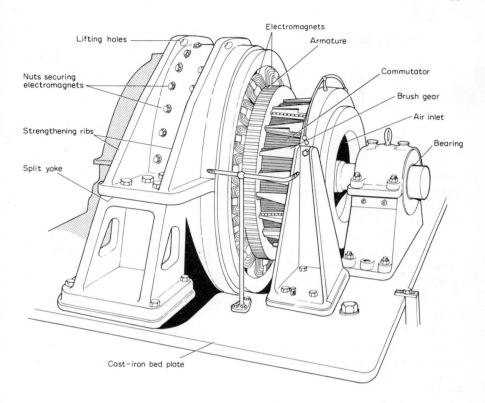

Lifting holes

Nuts securing
electromagnets

Strengthening ribs

Split yoke

Electromagnets

Armature

Commutator

Brush gear

Air inlet

Bearing

Cast-iron bed plate

Let us now consider a complete control system as used in practice. The system to be discussed is the Ward Leonard system of speed control as used on the continuous steel casting plant.

Ward Leonard Speed Control

In the method two identical direct-current machines are used as a motor and generator, and since these have not yet been discussed some introduction is necessary. A direct current motor and a direct current generator comprise the same basic machine but they are run in different ways. The motor is supplied with current to produce motion, whilst the generator supplies current when driven. The basic machine consists essentially of six parts: armature, commutator, brushgear, electromagnets, bearings and support frame. The armature carrying the coils, commutator and electrical interconnections are fixed to a shaft which is supported from and rotates in two end bearings. In large machines the bearings are of the pedestal type and are fixed to a solid bedplate which also carries a support frame for the electromagnets and brush gear. In small machines the bearings fit into end-shields locking into the

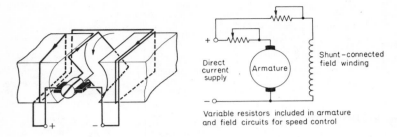

Shunt-connected direct current motor.

support frame. When used as a motor or a generator, current flowing in the electromagnets produces a magnetic field linking the armature conductors. When the machine is used as a motor, an increase in armature voltage leads to an increase in speed and an increase in field current leads to a decrease in speed. When the machine is used as a generator running at a constant speed an increase of field current leads to an increase of output voltage. (The commutator — to be discussed in more detail later — ensures that in the case of the motor the torque on the armature is always in the same direction and that, in the case of the generator, the output voltage always acts in the same direction.)

With these remarks in mind one can consider the Ward Leonard control system to be made up of two separate units; the first, a main direct current drive motor and the second, a direct current generator driven by some form of constant speed engine. The engine might be either of the diesel type or, more usually, an electrical three-phase induction motor. As mentioned before, such a three-phase motor is able to maintain a constant speed within a few per cent for all mechanical loads.

The field current for the driven direct current generator is jointly supplied from the direct current mains and the tachometer fitted on the end of the shaft of the main direct current drive motor. It should be noted that the direct current mains and tachometer voltages are connected in reverse. Under normal conditions the resultant voltage acting around this circuit is just sufficient to circulate the correct magnetizing field current. (A form of electronic current amplifier is also fitted but the details are omitted.) This generator field current then produces the correct generator output voltage to supply the armature of the main direct current drive motor — a sort of 'private' supply. The motor field coils are fed separately from the direct current mains. With steady mechanical loading the supplied armature voltage gives the right armature current to produce the desired motor output torque.

To see how the automatic speed control works, consider the following cases. Suppose the motor were working at a steady speed. If a sudden higher mechanical load were put on the motor it would tend to slow down and the tachometer voltage fall, but because the input mains voltage to the generator

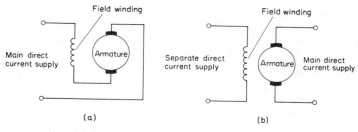

(a) Series-connected direct current motor.
(b) Separately-connected direct current motor.

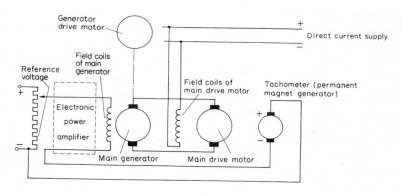

Ward Leonard constant speed
motor control system

field coils is constant the resulting voltage acting around the field circuit increases, thereby increasing the field current. (Remember that the mains voltage and the tachometer voltage are reverse connected so that decreasing one of them leads to an overall increase in the resultant voltage.) The direct current generator voltage would increase as would the speed of the drive motor since it is now receiving a bigger armature voltage. The speed will thus correct itself.

On the other hand, if the mechanical load were reduced or even completely removed, the speed of the driving motor would increase and the tachometer voltage would rise. The resultant voltage acting around the generator field circuit would be less than under normal conditions and the field current would fall. Hence the generated voltage would decrease and also the voltage supplied to the drive motor armature. The reduced voltage supplied to the motor armature would lower its speed. The speed has again corrected itself.

We have shown how the driving motor speed may be kept constant with varying mechanical loads such as those encountered in the continuous steel-casting process. In practice the three direct current motors driving the mould oscillator, withdrawal rolls and straightener rolls are supplied from one generator. (It was necessary, you remember, to have speed matching between all three.)

The main disadvantage of this form of automatic control relates to the need to have a separate motor-generator set of comparable size to the main drive motor. This, of course, involves unwanted losses and extra expense. In recent years, electronic devices called 'thyristors' have been developed in an alternative scheme, but although they are superior in many ways to the above method they also suffer from some disadvantages.

4. Computers and Automation

Historians generally agree that the Industrial Revolution began in Britain in the late eighteenth century, with mechanization in the cotton and iron industries and the development of the steam engine. Many pronounce that the revolution continues unabated, but others contend that there has been a sequence of revolutions. Following this latter approach one might argue that the first revolution, usually considered in traditional history books, was a Revolution in Machines. The Hargreaves 'Spinning Jenny' dates from 1764 and Arkwright patented his powered spinning frame in 1769. Smeaton made the first cast-iron blowing machine for an elementary blast furnace in 1760. By 1800 Maudslay had improved the lathe beyond recognition by incorporating inventions that greatly improved its cutting accuracy. Watt's development of the steam engine as a prime mover occurred at about this time and gradually replaced other prime movers such as the natural elements of wind and water, animal and human muscle power. Since that time other engines using electricity, petrol, diesel and other fuels have also been invented.

The second Industrial Revolution, running concurrent by the first, may be called the 'Control Revolution'. This revolution involved a change whereby the machine, process or device was operated by some external agency or control mechanism. A good example, discussed in the previous section, was the centrifugal speed governor on Watt's engine. Nowadays, control can be mechanical, (as in Watt's engine), hydraulic, pneumatic, electric or electronic or of some mixed arrangement. In general, however, electronic methods are preferred because they can be remotely operated and give precise control. In the early decades of the twentieth century applied mathematicians like Routh and Hurwitz began to lay the foundations for a generalized approach to all control systems. The great land-mark occurred in 1932 when H. Nyquist of the Bell Telephone Laboratories, U.S.A. presented a classic research paper on feedback.

He derived the criterion by which it is decided whether a closed-loop control system will be stable, conditionally stable or unstable. Today, most factories employ a variety of semi-automatic machines, processes and devices in their production lines using this sort of control.

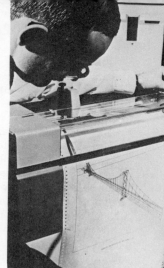

An example of a computer being used in a worldwide information network is B.O.A.C.'s Boadicea ticket reservation system operated from Boadicea House, London (above, left).
(By courtesy of B.O.A.C. and I.E.E.)

Computers used for design (above, right). Here a graph plotter has produced a perspective drawing of a bridge structure. From a mathematical model of the suspension bridge design, computer programmes are able to produce these perspective drawings from any viewpoint.
(By courtesy of I.E.E.)

Individual units of a production plant usually function automatically when set up, loaded and started. They will run for an indefinite time until their raw material runs out. Human intervention is only required at the beginning and end of the work programme, but some sort of general supervision is usually undertaken. True mass production, however, is only possible when it is practicable to control accuracy and reproducibility in manufacture. Mass production, or at least batch production, has been attempted since 1798, when Eli Whitney, the great popularizer of interchangeability, produced guns for the U.S. government. Not until well into the nineteenth century was true mass production possible with the invention of the turret lathe and milling machines. The beginnings of mechanized handling started about 1890 with flow production of small castings for railway brakes by the Westinghouse Company. The greatest credit for the development of mass production techniques using an assembly line must go to Henry Ford in his Detroit factories, who set out to provide 'motor cars for all'. In October, 1913 it required about ten hours labour time to assemble one motor; six months later by a moving assembly method the time was cut to about five hours.

Today we live at the beginning of a third type of revolution which may be called the 'Computer Revolution'. The speed of electronic computers as distinct from the earlier mechanical calculators and their modern electronic counterparts has opened a new era. It is well known that modern computers are ideal for routine clerical work in finance, commerce and industry. Jobs such as compiling bank records, maintaining customers' record cards and calculating net salaries immediately come to mind. It is less well known that computers are vitally important in science and technology, both in the theoretical and practical fields. Scientists and engineers are able to make calculations quickly and accurately with electronic desk computers, and in some cases solve problems that would have seemed impossible a few years ago because of their length or complexity. The computer is able to produce answers in a few minutes that would have taken an office full of skilled calculators years to complete. A great number of machines and complex installations in industry, mining, transportation, navigation, communications and for strategic use would be unthinkable without them. Applications in industry range from the control of screw-down rolls and flying shears in a steel mill, to the optimum selection and cutting of stock timber to minimize costs in furniture manufacturing. The machine tool industry is rapidly becoming computerized. For example, it is now feasible to machine the complex shape of a turbine blade with exactness and reproducibility. Robot coal-cutting and stowing machines in coal mines would be impracticable without the assistance of computers. In transportation, road traffic control, automatic trains on the London Underground, steering of large oil tankers and landing aids for aircraft all depend on computers. Using the United States Navy satellites in orbit 1000 km over the earth, a ship's position may be quickly calculated to within 0.4 km in any weather. Navigation for the Q.E.2 and Concorde with the minimum of delay and maximum efficiency would be impossible without fast working computers. In a slightly different application, military combat missiles incorporate computers on board to enable them to track down and destroy their enemy objectives. In all these applications human operators would be far too slow to respond to the situation and to take the right decision.

Naturally, computers come in different sizes and styles and some work faster than others. A would-be buyer has a fairly wide range from which to choose. It is often more economical to buy time on a larger computer than to buy a machine. As in other hiring schemes, a subscriber pays a certain amount of rent per month plus the cost of the machine time actually used. Access to a computer by a subscriber using a time-sharing arrangement with other subscribers is achieved by each participant having a 'terminal' on his premises. All such terminals are electrically connected to the main computer installation and switched in and out of operation as the work programme and financing allows.

Computers may be divided into digital or analogue types. The digital computer (see Appendix VIII) is the most expensive and versatile type, and works by counting as in arithmetic. It is able to perform the simple processes of addition, subtraction, multiplication and division. Other operations can be performed in special ways and might, for example, involve the evaluation of a mathematical function such as a sine or cosine of a number. The

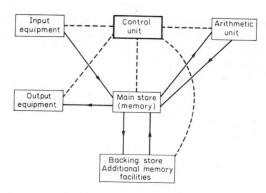

Block diagram of a digital computer.

The control unit links and coordinates all the basic parts of a digital computer. Two of these parts are the main store or memory and the arithmetic unit. The main store stores instructions and data and the arithmetic unit does all the calculations.

In operation the control unit accepts the machine instructions in sequence and decodes them. Data and decoded instructions are fed into the arithmetic unit and the necessary computation performed. After completion the answer is transferred back to a storage section in the main store. If further computations are required information is again transferred to the arithmetic unit and the new calculation performed. The final answer is then fed back to the main store and hence to the output equipment.

value of such a function can be obtained by the well known mathematical method of series expansions and summing of sufficient terms to give the required degree of accuracy. Commonly used mathematical functions such as sines or cosines are usually recorded on a special tape, wire or card and referred to as 'sub-routines'. They are called into action whenever 'spelt out' in the programme.

Digital computers are often mistakenly called 'electronic brains', but they are certainly not thinking machines in the sense of being capable of original thought. They are basically fast and efficient calculating machines, possessing superb memories and the ability to accept huge amounts of information, to make decisions on courses of action, and from various deductions to give commands. (A computer is only as good as its designer and programmer in that it simply follows instructions.) To produce numerical answers to any specific problem three types of information must be fed in: (i) input and output instructions (ii) programme of work or calculation to be performed and (iii) data or figures to be used in the calculations.

The analogue computer is not so revolutionary as the digital computer in the 'Computer Revolution' since it really forms a logical extension of control system theory using simulation techniques. It is usually bought with a specific application in mind. The problem to be solved has its different quantities represented by analogous quantities in electric circuits, and the analogue computer (see Appendix IX) solves its problems not by counting as in a digital computer, but by measurement. Such a computer has an inaccuracy of about 0.1 per cent but this is good enough for most design purposes. On the other hand, a digital computer can have an accuracy as high as required. A typical version is able to work to eight significant figures or, if required, with double precision to sixteen significant figures.

One of the most useful applications of the analogue computer is in the design of motor car shock absorbers, the ideal absorber would be one that would absorb the initial impact as smoothly as possible and reduce subsequent vehicle bounce to a minimum. Here a choice must be made between helical springs of various strengths and between types of damping fluids. The production of a good practical design using real components would be long, tedious and cost a good deal of money, as a variety of combinations of springs and fluids would have to be manufactured, fitted and tested in turn. If an analogue computer is set up to represent the arrangement, the optimum design will be arrived at quickly. The various vibration patterns can be either plotted with pen recorders, or better still, displayed immediately on an oscilloscope. A further advantage of such computer-aided design is that any of the variables is continuously adjustable, not simply changed in a step-by-step sequence. Thus there is the possibility of detecting any unsuspected natural resonances in the structure.

Another interesting example of the analogue computer is its employment in a flight simulation training rig. The trainee pilot sits in a mock flight deck similar in most respects to the one on an actual aircraft, and all the necessary manoeuvres for take-off, flight and landing may be simulated — without there being danger to aircraft, property or personnel. Various electro-mechanical transducers respond to the trainee's reactions and feed data in the form of various voltages to the analogue computer. The computer immediately responds with output signals which cause other electro-mechanical transducers to produce associated physical motions of the flight deck rig. Related changes in instrument readings also occur. In other words the flight deck responds accurately to the pilot's actions. Variations of air noise with speed, engine pitch with throttle and even tyre screech on touch-down are all simulated. A pilot can be trained for abnormal as well as normal flight conditions. Engine overheating, fuel failure and many other engine and airframe malfunctions can all be reproduced. A variety of situations can be presented to the trainee and his reactions checked. Naturally, in most cases, the emergency situation could be directly produced in an actual aircraft but this is expensive and dangerous. The advantage of the computer-controlled flight simulator is that there is no risk. It is cheap and a complete catalogue of normal flight, and emergency situations can be presented to the trainee in a few minutes. Flight training simulators are available for most big aircraft currently used for civil and military purposes — including, of course, Concorde and the Jumbo jets. In fact the first Concorde pilots were trained on simulators long before the planes were built.

Turning now to the use of a digital computer, its speed and precision are probably particularly suited to applications in aerospace work. Jet airliners need miniature computers to aid their navigation because the mental capabilities of a human being are just not good enough to calculate the minute-by-minute course corrections necessary to ensure the following of a precise flight plan. Extra-terrestrial exploration such as the Apollo manned lunar space mission, would be unthinkable without a computer. Huge computers are situated at the Mission Control Headquarters and, of course, the space vehicle modules have their own miniature computers on board. Accurate navigation in the lunar landing and earth re-entry and their monitoring are particularly important in these space missions. In the case of re-entry, high-flying aircraft and a large number of land-based radio stations pick up information about the position and timing of the space vehicle during the period. This data is then fed back to the Mission Control computers where the trajectory is plotted and the exact point of splashdown predicted. The recovery team in their helicopters are then able to 'sprint' to this spot with valuable minutes saved — needless to say, it could be a matter of life or death.

An earth-bound example of the use of the digital computer is its application in power generating stations. Recently, the 2000 megawatt plant at Fawley became operational and is one of the most advanced in the world because of its high degree of computerization. Each of the four boiler/turbine units are controlled by English Electric KDF7 computers which allow automatic starting-up, loading, running to maximum efficiency and shutting down. The computer system is able to give alarm analysis and take remedial action on any plant failure that may occur. Further information on the type of fault and what action has been taken or should be taken is displayed on a cathode ray oscilloscope. The wide range of plant under control of the computer system includes such diverse equipment as the fuel-oil system, the 32 oil-burners, the boiler feed pumps, lubrication arrangements, oil-seal systems, steam emergency stop valves, electrical synchronizing equipment and the generator step-up transformer to the Grid. The computer receives information from 1200 analogue and 1800 digital inputs on each of the four boiler/turbine units and uses this data to control approximately 350 items of plant. To appreciate the complexity of the problem it might finally be mentioned that the computer programme has over 1 million 'words'. In 1971, Wylfa, the twin-nuclear reactor station in Anglesey, went on load for the first time with 1000 megawatt. It was the first of the nuclear stations to be designed for central computer control.

Finally, in retrospect, how do all these computer applications affect our overall thinking of the use and possibilities of electricity in factories and elsewhere in the future. One feature is quite clear, electricity and electronics provide the means for automation at a whole range of different levels. Small computers designed for individual machines as well as with major computer installations, with or without time-shared arrangements, will become increasingly common. Over the years, small computers will cease to be considered as a separate entity and will instead be seen as a component for incorporation into a total system. The big computers enable the inputs and outputs of different semi-automatic machines in a factory production line to be related. This enables total plant optimization so that the maximum production can be achieved with maximum efficiency. Even today, at this early stage of computerization, several major oil and chemical industries plan their activities with the aid of a computer on a day-to-day basis to maximize profits. There is, of course the further possibility, however remote, of factories being completely run by computers. In the extreme limit, computerization would involve organization and management as much as production plant. The advent of automation, or cybernation as the experts prefer to call it, as distinct from mechanization and its elementary control, is engendering a revolution in our society as important as the changes that occurred in the original Industrial Revolution.

Costing
Before completing this chapter on electricity utilization, let us look at a typical engineering problem where an important assessment and decision are made.

British Rail Electrification
In the 1960s at considerable expense, a large section of British Rail was electrified between London and Manchester and Liverpool — about 800 route km. The system chosen was the 25 kilovolt, 50 hertz, alternating current arrangement, using overhead conductors to feed energy to the locomotives.

Power is fed to the overhead conductor system by a number of transformer substations located along the track. Pantographs, on top of the locomotive, pick up the current. After collection by the locomotive the power goes through a circuit-breaker, transformer, and bank of semiconductor rectifiers, where it is converted to direct current. This is supplied to d.c. series-connected motors driving the axles of the locomotive. The reasons for the choice of the 25 kilovolt alternating current as opposed to other systems, was based solely on the fact that it was cheaper than the other possibilities.

The main alternative was the 1.5 kilovolt direct current system used, for example, on the Manchester-Sheffield main line. This method employs a relatively large number of transformer/rectifier substations feeding the overhead conductors and spaced along the track. At these substations the mains alternating current is converted to direct current, and the pantographs collect the power which is fed directly through controllers to the usual d.c. series motors.

Let us compare the two systems assuming that two cities are 320 km apart and are to be connected by a double track system. If the scheme requires 16 locomotives and 25 three-car units, the costings would be approximately as follows:

25 kilovolt alternating current system

640 km of track of overhead line at £6500 per km (320 route km × 2)	£4 160 000
3 substations at £80 000	£240 000
25 three-car units at £110 000	£2 750 000
16 locomotives at £80 000	£1 280 000
	£8 430 000

1.5 kilovolt direct current

640 km of track of overhead line at £9000 per km (320 route km × 2)	£5 760 000
18 substations at £85 000	£1 530 000
25 three-car units at £100 000	£2 500 000
16 locomotives at £70 000	£1 120 000
	£10 910 000

As can be seen from this first estimate, the alternating current system would appear the cheaper. This is not necessarily the case however, since the apparent saving might be easily swamped by the cost of civil engineering work in order to provide the larger electrical clearances. The increased speed and traffic density also usually involves new signalling and track arrangements. In other words, simple calculations can be sometimes misleading, one needs to know all the costs. Besides, comparisons are only valid when the total price includes running over an estimated life, as well as capital costs.

5 The Birth of Electricity

We have seen how electricity is generated, transmitted, distributed and utilized in the modern industrial world. Let us now see how electricity supply came about; to do this, of course, we must go back in history.

The early origins of electrical science can be attributed to the Ancient Greeks who were aware of a few simple electrostatic phenomena and of the magnetic properties of naturally occuring 'lodestone'. Authentic records describing the use of lodestone as a navigational aid date from about 1000 A.D., but little true scientific investigation occurred before 1600. In that year William Gibert published his great work on basic magnetism and static electricity, and interest developed in the eighteenth century with speculation on the nature of lighting and means of protection against it. The real impetus to electricity was due to Professor Galvani who discovered that two dissimilar metals in contact with a frog's leg caused it to twitch. Subsequently, in 1796, Volta invented the first electrical storage battery, which consisted of alternate metals separated by cloth or paper soaked in an acid electrolyte. Frictional electric generators and 'influence' machines relying on electrostatic principles had also been invented by this time but were not reliable and only had a small output. In fact when our story really begins, the only adequate sources of electricity were the recently developed chemical cells.

In 1809 Humphry Davy used a huge battery of cells to give the first spectacular demonstration of the use of electricity to produce light. At a lecture to the Royal Institution he passed a large current between two carbon rods to produce an intensely bright arc. The experimental lamp was a great success and aroused much interest but it was not until the middle of the nineteenth century, with the invention of suitable generators, that arc illumination could be brought into general use.

In 1812 the most enthusiastic member of Sir Humphry's audience was a young man named Michael Faraday. He had read as much as he could on the subject and was held spellbound by the lecture. At the end he copied down all his notes, bound them with his own hands — he was an apprenticed book-binder at the time — and presented them to Davy. Sir Humphry was most impressed at his understanding and the kindness of the gift, and it was through Sir Humphry's recommendation that Faraday was able to make his hobby into his job and join the laboratory staff at the Royal Institution.

Throughout his career, one of Faraday's main interests was the connection between electricity and magnetism. In 1820 Hans Christian Oersted, a Danish physicist, who like Faraday was interested in electricity and magnetism, made a remarkable discovery. His results were published in a small pamphlet written in Latin which fortunately he sent to a large number of scientific societies. Oersted discovered that when a magnetic compass needle was placed near a conductor carrying current from a voltaic cell the needle was deflected, implying that the current-carrying conductor was surrounded by a magnetic field. This relatively simple observation was of momentous significance and immediately earned him international honours.

Oersted's experiment was repeated and confirmed in many laboratories and, in particular, a Monsieur André Marie Ampère of Paris became very interested. Within two months of the publication of Oersted's paper, Ampère, who had a keen mind and was a first class mathematician, had produced further important results which showed that the direction of current flow in the conductor determined the direction of the compass needle deflection. Further, he argued that if the electric current could produce a force on a compass needle, then two conductors carrying currents should produce forces on each other by virtue of their respective magnetic fields. If the current-carrying conductors lay parallel, their magnetic fields should produce mutual attraction or repulsion depending on relative current flow directions. With this insight and interpretation he provided the basis for the invention of the practical electromagnet and eventually the electric motor.

Faraday took up the thread. If electricity could produce magnetism, then surely magnetism should produce electricity. Nature must be symmetrical. Although occupied with research work in chemistry and electro-chemistry during the following years, Faraday's thoughts on this problem were always at the back of his mind. For example, in 1822, there is a short note in one of his laboratory notebooks — 'Convert magnetism into electricity'! The crucial steps in achieving this end were taken in the late summer of 1831. Electricity could be made from magnetism.

The Experiments of Faraday

Faraday plunged a bar magnet into a circular coil of wire connected to a simple detector — developed from Oersted's findings — and generated a momentary electric current. When the magnet stopped then the current stopped, and when the bar magnet was pulled out of the coil, current flowed in the opposite direction. In Faraday's own words 'Every time the magnet was put in or pulled out, a wave of electricity was produced'. Furthermore, if the coil was moved with the magnet held stationary, the same effects were observed. Hence Faraday concluded that it was only when *relative motion* occurred between magnet and coil that electricity was generated. In November 1831 he demonstrated the continuous generation of electric current using magnetism. The apparatus consisted of a large copper disc revolving in the gap

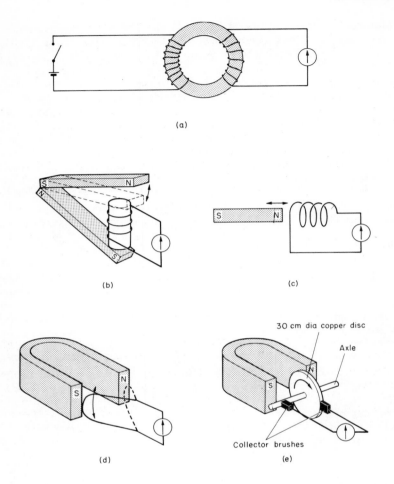

(a)

(b)   (c)

30 cm dia copper disc

Axle

Collector brushes

(d)   (e)

between the poles of a big permanent horseshoe magnet, the current being collected from the elementary generator by contacts sliding on the outside edge of the disc and on the axle. The faster the disc was turned, the greater was the current output. The mechanical energy put into turning the disc was converted into a new form of energy — electrical energy.

In the same year, Faraday also discovered the principle of the transformer, which consisted of two coils of wire wound on a wooden- and later on an iron-core but not electrically connected. (Faraday probably used milliner's wire which was normally employed for stiffening ladies' hats, as this was the only type of wire available at that time. Electrical insulation to prevent a short-circuit was a problem; Faraday wound a piece of twine between

The experiments of Faraday (opposite). Faraday showed that magnetism could be used to generate electricity.

(a) He discovered that when a current in a coil of wire, wound on an iron core, was switched on and off a *momentary* current was induced in a neighbouring but not electrically connected coil, wound on the same core.

(b) A similar induction effect occurred to a single coil placed in a magnetic circuit formed by an iron core and a pair of magnets. When one of the magnets was moved to change the circuit a surge of current was sent through the coil.

(c) The most famous demonstration of electromagnetic induction was when a magnet was thrust into or out of a stationary coil. If the magnet was held stationary and the coil was moved, current again occurred. When no relative motion took place, no current flowed in the coil.

(d) A loop of wire moving up and down between the poles of a magnet gave the same transient effects.

(e) Finally, replacing the two conductors in the previous experiment by a complete copper disc which was rotated between the poles of the magnet enabled a continuous current to be produced. The current was collected from contacts sliding on the axle and the edge of the disc.

consecutive turns of wire and placed calico between layers.) When a steady current was passed through one of the coils, the second coil was unaffected, but when the current was changed in one coil, an electric current was set up in the second coil. The theory put forward was that the changing current in the first coil produced a changing magnetic field or flux which also passed through the second coil, and it was this changing flux through the second coil that generated the current. On 24 November 1831, Faraday summarized all the results of these experiments in the most important lecture that the Royal Institution has ever held.

Practical Generators

The achievements of Faraday soon led to the production of practical electric generators. Some used a conductor moving relative to a magnet as in Faraday's machine; others used a magnet moving relative to a conductor. One of the earliest types, really a large 'scientific toy' by modern standards was devised by Hyppolyte Pixii, probably an Italian and exhibited in Paris in 1832. Although mainly used for demonstration purposes, it was also used to give electric shock treatment to paying volunteers suffering from rheumatism who believed in its therapeutic value.

The input mechanical power to Pixii's machine was used to rotate a horseshoe magnet past two stationary coils. As the magnet poles swept past the coils, electric current was induced. The two coils were connected so that their

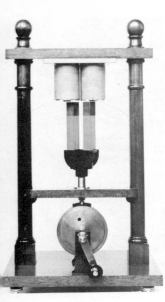

A Science Museum replica of Monsieur Pixii's hand-driven permanent magnet generator (left). The original machine was constructed at the end of 1831, after the publication of Faraday's researches in electromagnetic induction. It is an alternating current generator but a later version, incorporating a commutator, produced direct current.
(By courtesy of Science Museum.)

Pixii's magneto-electric machine and commutator (below).

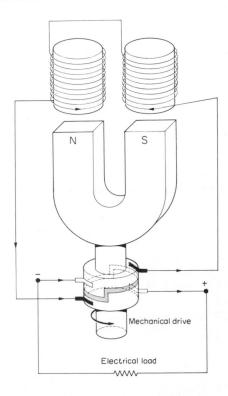

currents added. Electrical power output was small because the machine was hand-driven and the current produced depended mainly on the physical strength of the operator! Further, as alternate poles of the magnet swept across each coil, the direction of current in each coil changed. Current flow was first in one direction and then in the other; such a current is said to be alternating.

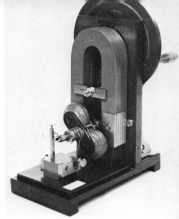

Magneto-electric generator due to E. M. Clarke (about 1834).
These early commercially produced machines were used in many
laboratories to provide electric shocks for various physiological
experiments. They were also used in Victorian households for
their supposed ability to cure rheumatism and other ailments.
The Clarke generator was similar in principle to the Pixii generator
but differs in that the permanent magnet is kept stationary whilst
the bobbins rotate.
(By courtesy of Science Museum.)

Later at the suggestion of Ampère, Monsieur Pixii fitted a reversing switch
or commutator to the machine. The commutator was geared to the main
shaft and made to operate automatically as the horseshoe magnet rotated. Its
working can be explained as follows. When the south pole of the magnet came
under the coil it appeared as a south pole at the output terminals. However,
when the magnet and shaft rotated so that the north pole came under the
coil, the commutator reversed so that it appeared as if south pole were
coming. Subsequently, when the south pole came under the coil, the commutator was switched back to normal. Hence a succession of south poles
appeared at the output terminals and a unidirectional flow of current was
obtained. Such a combined arrangement of magnet, coil and commutator is
called a direct current generator.

From this early date direct current generators were more popular than the
simpler alternating current generators because the unidirectional current flow
was similar to that obtained from chemical storage batteries. Direct current
was thought to be more readily understood, controlled and useful. Today
many engineers claim that the invention of the commutator retarded the
progress of electrical machines for fifty years, because alternating current
generators have far more technical advantages.

Other inventors besides Pixii tried their hand at making more efficient
generators but none were very successful. A basic problem of developing
machines at this time was that it was difficult to judge performance. The sole
means of assessment between two generators was which produced the biggest
shocks or sparks. For example, Saxton's machine of 1833 generated bigger
shocks than Pixii's and it was therefore thought to be superior, but the only
reason it produced bigger sparks was because of the larger amount of wire
wound on it. Twenty years were to elapse before there was a real break-
through.

The delay may be attributed to the fact that few inventors and scientists
really understood electromagnetic induction. The concepts of magnetic
field lines and the cutting of the lines, introduced by Faraday, were
difficult to understand. The best arrangement of magnetic field and coil conductors to give the highest current output does not seem to have been regarded as a problem, bigger outputs being obtained by having more magnets
and connecting in more coils. Secondly, the ways in which electricity could

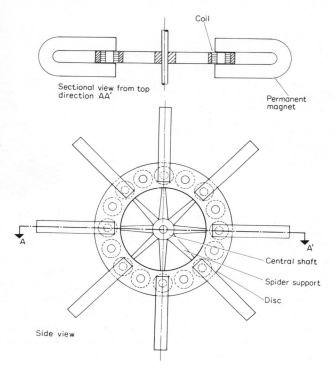

Coil

Sectional view from top
direction AA'

Permanent
magnet

Central shaft

Spider support

Disc

Side view

The construction of an early magneto-electric generator,
1851 (above). (By courtesy of I.E.E.)

Professor Holmes direct current magneto-electric generator
designed for the Dungeness lighthouse, 1862 (below).
   (By courtesy of the Institution of Civil Engineers.)

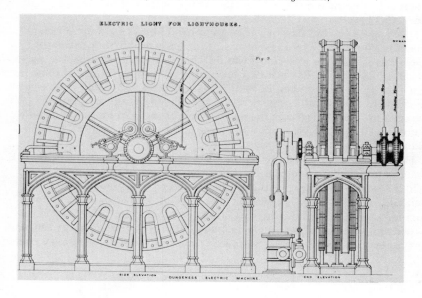

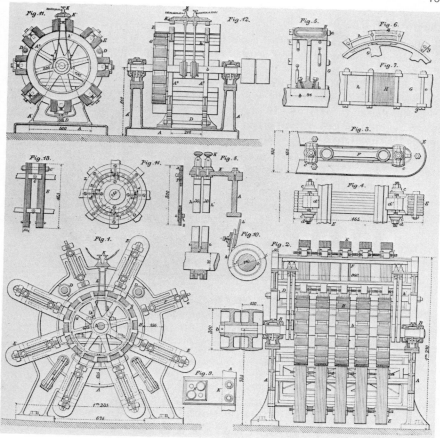

De Meritens magneto-electric generator — 1881 — designed at least
fifteen years after the introduction of dynamo-electric machines.
(By courtesy of 'Engineering'.)

be applied were limited and no great incentive existed for improving the
machinery. A few electric arc lamps were used in public buildings and a
number of lifting electromagnets were used in steelyards, but improved
chemical storage batteries supplied these fairly satisfactorily. Only when the
increasing popularity of electricity gave rise to the demand for bigger, more
reliable and convenient power supplies did generator development really
commence. In the 1880s this impetus was to be provided with the invention
and widespread use of the carbon filament lamp. Apart from the addition of
a steam-engine drive the permanent magnet machine in its basic form remained
in use for a long time.

Electric Motor Action

Parallel with the development of the electric generator lay the development of the electric motor. By 1833 several inventors had produced primitive types of motors, and by the end of the decade large versions were being made. For example, in 1842 Robert Davidson built a battery-powered electric locomotive weighing 5000 kilogramme which propelled itself on an experimental run from Edinburgh to Glasgow. The project ended in disaster when a mob of steam locomotive stokers and drivers destroyed the engine because they claimed it threatened their jobs.

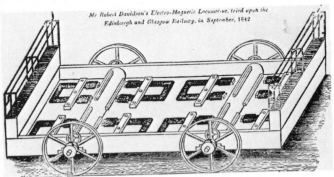

Davidson's electromagnet locomotive (above).
(By courtesy of Science Museum.)

Early poster advertising an exhibition of Davidson's locomotive (below).
(By courtesy of Science Museum.)

(At this time J. P. Joule, who enunciated the First Law of Thermodynamics which related mechanical work and heat, compared the costs of electrical and steam power and concluded the latter was the cheaper by a large factor.)

Many of these primitive motors were modelled on the action of the then fashionable reciprocating steam engine. Typical of these was one invented by Professor Page from the United States. It was based on the well known double-acting beam engine, the steam cylinders being replaced by large electric solenoids and the pistons by iron cores connected to opposite ends of the pivoted beam. The solenoids were energized alternately, pulling down their respective iron cores and causing the beam to rock; the rocking motion was converted to rotary motion by means of connecting rods and a large flywheel. The maximum output of this machine was about 1 kilowatt but a much larger type built for traction purposes in 1854 achieved the incredible speed of 30 kilometre per hour with a load of over 12 000 kilogramme. The experimental machine suffered many subsequent mishaps and the U.S. Congress, which financed the project, withdrew its support.

About the same time as Faraday's researches, D. F. J. Arago, famed for his researches in optics, made the first observations of an elementary motor action which has become the basis of nearly all twentieth century motors. It is worth considering this in some detail, because of its subsequent importance.

Another electric steam engine due to Bourbouze.
(By courtesy of Burndy Library.)

Arago noticed that if a magnetic compass needle was set oscillating it continued to do so for some time, but that if a copper plate was close to the needle the oscillations quickly stopped. Arago concluded that some unknown force must be present. This force only manifested itself when *relative motion* occurred between the magnetic field and the copper conductor. By arguments of symmetry similar to Faraday's, Arago suggested that the same force must be set up when the copper conductor moved relative to a stationary magnetic field, and he devised an experiment to test this hypothesis. The apparatus comprised a hand-driven copper disc rotating immediately below a magnetic compass needle and, as he anticipated, the latter was deflected when the disc revolved. The faster the rotation, the greater was the deflection, and when the disc was driven very fast the needle revolved. When the disc rotation was reversed the needle reversed, i.e. the needle always tended to move in the direction of disc rotation. Arago did not fully understand the significance of these discoveries but the meaning became clearer after later experimental results (see Appendix X).

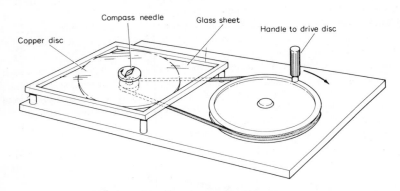

Arago's Disc (above). When the copper disc was rotated it dragged the nearby compass needle around with it.

Producing a rotating magnetic field (below).

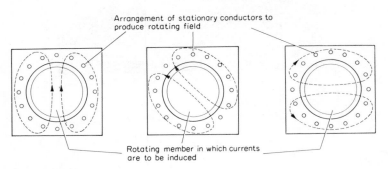

The converse of the Arago experiment, in which the roles of compass and disc are interchanged, also gives motor action. Thus, when a magnetic field due to a compass needle or some other source is caused to rotate about some axis, it will drag around with it a smaller copper disc. The development of these principles into a more suitable form eventually led to the production of the commercial induction motor, but for more than 50 years the phenomenon remained a curiosity. The biggest problem was the production of a good rotating magnetic field.

A later development of the Arago experiment occurred in 1879 due to Professor W. Bailey. He replaced the single magnet by four electromagnets spaced 90° from each other below a copper disc which was free to rotate. The electromagnet coils were fed from a battery via a hand-operated rotary switch. When the switch handle was turned the current was switched to change the polarity of the coils, adjacent coils having opposite polarity. A rotating field was produced and this caused the disc to move. Professor Bailey regarded the apparatus as just another scientific 'toy' having no commercial possibilities.

In 1888 the rotating field problem was solved independently by Galileo Ferraris of Turin and Nikola Tesla of Smiljan, a small village in what is now Yugoslavia. Ferraris and Tesla obtained the necessary action by placing the electromagnets at right angles and passing through them alternating currents which reached their maximum values at slightly different instants of time — one being a quarter of a cycle behind the other, i.e. having an electrical phase difference of 90°.

Like his predecessor Ferraris predicted that the apparatus would have no future as a powerful motor but that the operating principle might be useful

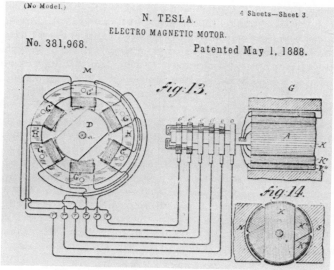

Nikola Tesla's three-phase motor and generator.
(By courtesy of I.E.E.)

Experimental single-phase induction motor. In this motor made
in 1891 by W. Langdon-Davies the difficulty of starting a single-
phase motor from rest was overcome by a method of 'phase
splitting' which gave the necessary rotating magnetic field. A
starting winding was in parallel with the main winding, a non-
inductive resistance being included in one of the circuits so that
the necessary phase difference between the currents in each
circuit was obtained.
(By courtesy of Science Museum.)

in an electricity meter. Tesla, however, had more foresight and perseverance.
By means of a succession of brilliant inventions in the late 1880s and early
1890s, leading to a total of thirty-seven patents, he established the possibility
and practicability of obtaining motors without commutators. Amongst these
discoveries was a new generator coil arrangement called a three-phase system.
In the method the output was delivered from three wires whose currents had
a phase difference of one third of a cycle or 120 electrical degrees. When this
supply was fed to a similar coil arrangement in the stator of a three-phase
induction motor, the rotor of the motor turned smoothly with great power.
George Westinghouse in the U.S.A. secured Tesla's patent rights but it was a
few years before full production commenced in 1892. Tesla's patent rights
were infringed on a wide scale; in fact two years earlier a 75 kilowatt three-
phase induction motor was demonstrated at the Frankfurt Exhibition.
Although the induction motor was quite unsuitable for traction purposes —
there was a ready market at the time — it was ideal for factory drives.

A long time elapsed before it was realized that an electric motor was the
converse of a generator. A generator converted input mechanical energy to
output electrical energy, whereas a motor converted input electrical energy
to output mechanical energy. In 1860 an Italian physicist, Dr. Pacinotti, who
afterwards became a Professor at Pisa University, showed that a generator
would work equally well as a motor. Once the nature of this relationship was
verified the technology developed for the generator could be applied to the
motor.

Langdon-Davies single-phase induction motor. This is a commercial form of the Langdon-Davies experimental motor, patented in 1894—7. The stator and the rotor of the motor have been partly sectioned.

There are two sets of stator windings, one of which is used for starting only and is angularly displaced with respect to the other set. The stator core is built up of stampings perforated to receive these windings. A starting switch is used for introducing and cutting out the starting winding. The rotor consists of a series of bars which are arranged in slots and connected at the ends by metal rings. For better starting these bars are arranged askew.

(By courtesy of Science Museum.)

6 Electricity Grows Up

The initial demand for electricity in bulk arose in the middle of the nineteenth century for electroplating and the processes used in printing. Both industries originally employed batteries of electrochemical cells to supply the current but as demand increased batteries were unable to provide the power required. In 1842 John Woolrich devised a big permanent magnet generator to supply current for electroplating purposes in his Birmingham factory. Soon gold- and silver-plated ornaments and table utensils were being mass produced and brought within the price range of a larger section of the public. At the same time the electric telegraph method of communication was being developed, especially for the new 'railways'. Its power requirements were small but it did a lot to make people familiar with electrical techniques. The telegraph line and its possibilities attracted enormous publicity in 1845 when a suspected murderer was captured at Paddington, the news of his expected arrival being telegraphed from Slough where he was seen to board the train.

Arc Lighting

The greatest impetus to the development of electric generators was due to the lighting industry. In particular public interest had been stirred by Grace Darling and her father who rescued the passengers and the crew of the steamer 'Forfarshire' which foundered on a stormy night off the Northumberland coast. This drew attention to the need for providing better lighthouses to warn mariners of dangerous rocks. In 1858 the first installation of an electric arc light powered by a generator in a lighthouse was completed at South Foreland off Dover in the busy English Channel. The system was devised by Professor Frederick Hale Holmes, who had recently taken out certain patents for improving the generators. Two machine-sets were supplied, one as a standby for maintenance purposes and in case of breakdown. Each set consisted of a reciprocating steam engine driving a direct current generator of about 2 kilowatt. The steam engine and the generator were directly coupled and ran at a speed of 90 revolutions per minute. Although each generator had 60 magnets, each weighing 100 kilogramme, and 80 coils, it was still no more than an upgraded version of the earlier permanent magnet or magneto-electric generators. The arc light had a luminous intensity of over 1000 candela and an automatic feed control for the carbon rods. Faraday, who acted as

Magneto-electric generator (above, right). This was one of the earliest generators used for lighthouse illumination and was installed in Souter Point Lighthouse near Sunderland in 1871, where it remained in use for many years. It was designed in 1867 by Professor F. H. Holmes whose work was largely responsible for the introduction of the arc lamp in lighthouses.

The field system consists of 56 permanent magnets mounted on stationary discs between which are placed 6 rotating armature discs carrying a total of 96 coils. The electrical output was 2.4 kilowatt collected from the armature by sliprings. A reciprocating steam engine running at 400 revolutions per minute provided the mechanical drive.

(By courtesy of Science Museum.)

Holmes arc lamp for lighthouses (1862) (above, left). This lamp, which worked for nineteen years, was constructed by F. H. Holmes and was probably the first electric arc lamp ever to be employed in a lighthouse. In the arc lamp the arc length was kept constant as the carbons burnt away by a simple control mechanism readily understood by old sailors who manned the lighthouses. The lamp also contained a means of focusing the light very exactly.

(By courtesy of Science Museum.)

technical adviser for the Trinity House Corporation, was much impressed by the power, stability, and ease of maintenance of the light, and presented a very favourable report. As a result further installations were ordered for Dungeness and Souter Point, near Sunderland. The Souter Point equipment with modifications was particularly successful and ran continuously from 1871—1900.

Arc lamp illumination in L'avenue de L'Opera.
(By courtesy of Science Museum.)

A typical use of arc lamps in the 1860s was to provide illumination for outdoor events, but it was not until the following decade that lighting of public streets and buildings was introduced. Successful installations were carried out in the Avenue de l'Opera in Paris and on the Embankment in London. Lamps were also used at the Gare du Nord, Paris and St. Petersberg dockyard, Russia. Billingsgate Market and the commercial ironworks of Wells and Co., both in London, were also electrified. In 1878 the first floodlit football match was held at Sheffield in front of 30 000 spectators. More British installations were completed in 1879 at the British Museum, Will's tobacco factory at Bristol and London Bridge railway station.

Portable generating set.
(By courtesy of Science Museum.)

The arc lamp installation at St. Enoch's railway station, Glasgow, used six arc lamps of 6000 candela, supplied by six Gramme generators to replace the original installation of over 400 gas jets.

Scientific travelling showmen advertized and hired out their portable sets to public works contractors wishing to work at night, and to public entertainments such as dances and concerts in the summer and ice skating parties in the winter. One of the most important of these 'electricians of the road' was R. E. B. Crompton. It was he who was later responsible for a major lighting installation at a big theatre in Vienna. The theatre was previously lit by gas but in 1883 was destroyed by fire with a great loss of life. For safety's sake Emperor Franz Joseph decided that the theatre should be lit in future by electricity. From 1885 to 1889 Crompton spent much of his time lighting the Opera House, theatres and other important public buildings, but for the sake of accuracy it should be pointed out that by this time a mixture of arc lights and vacuum filament lamps were employed.

Such examples indicate that lighting of this form was most suitable for the illumination of large public areas. Arc lamps were not really suitable for domestic installations because of the brilliance of the light, difficulties of maintenance, and high costs; hence gas lighting still held its own. Nevertheless by 1881 the number of arc lights in England alone exceeded 4000. Although the arc lamp retained its popularity as a form of street lighting until the beginning of the twentieth century, its future use was limited by the introduction of the carbon filament light attributed to Joseph Wilson Swan of England and Thomas Alva Edison of the United States. The arc lamp is still in use in cinema projectors and theatre spotlights where an intense point source of light is required.

Vacuum Filament Lamp

Swan began his experiments in 1848 and produced fine carbon filaments of good flexibility and high mechanical strength. These were mounted in glass bottles, evacuated by a simple handpump, and sealed off, leaving two wires for the current to pass in and out. Although he succeeded in rendering the filament incandescent, he abandoned the experiments because it quickly burnt out. In 1877, with the introduction of the Sprengel vacuum pump and hence the possibility of achieving a better vacuum, Swan resumed his experiments, but even with the better evacuation the filament still vaporized after a short time. He quickly realized the solution. The bulb should be exhausted of air and sealed off with the filament hot, otherwise gases trapped on the surface of the filament would be released when the filament was heated, thereby destroying the high vacuum so painstakingly achieved. Before long he succeeded in producing working lamps, and in 1880 he lit his own shop and the whole of his street with the new illumination. Before the end of 1880 an important industrialist adopted the new type of light at his country residence and the electric power was obtained from the first primitive hydro-electric plant.

Kensington Court power station (opened 1886–7). The power
plant comprised seven steam engines each driving an electrical
generator, giving the station a total capacity of 550 kilowatt. The
station and its specially designed distribution network mark the
beginning of modern electricity supply.

(By courtesy of Science Museum.)

Edison was an inventive genius with a good head for business, and as a former telegraph operator he had acquired some knowledge of electricity. In 1876 he used the proceeds of earlier inventions to set up his own private laboratory in New Jersey, U.S.A., where he studied the possibility of making electric lamps. At first he used metal filaments but later turned to thin layers of carbon. Amongst the materials he tried as substrates were paper — as in Swan's early experiments — horsehair, bamboo and silk sewing thread. He read a research paper, published in August 1879, on gases trapped on metals and quickly jumped to the same conclusion as Swan, that lamps must be evacuated whilst hot. In November of the same year he achieved succes and applied for British patents for his lamps.

Both Swan and Edison formed companies to market their inventions and were so successful that in the first few months, lamp prices dropped from more than a pound each to twenty-five pence (then five shillings). By 1882 both firms had expanded greatly, and Edison's company tried to secure an injunction to prevent Swan producing lamps, claiming an infringement of patent rights. Fortunately, both companies realized their interests were best served by cooperation rather than competition, and an expensive legal battle was avoided when the firms negotiated a private settlement and amalgamated, forming the Edison and Swan United Electric Co. Ltd. The joint company had the monopoly of manufacture in Britain until 1893, when the lamp patents expired.

One or two further remarks must be made about Edison, for he was not solely concerned with producing a cheap and serviceable source of light. He wanted to build a complete electrical distribution system in direct competition with the gas companies. To this end he built the Pearl Street Power Station in New York and its associated feeder lines in 1882. By 1884 he was supplying electricity to nearly 60 000 lamps. With his understanding of human psychology and the skilled use of advertising, Edison marketed his product very effectively. At the same time as showing that electric filament lighting was superior to gas in offering greater convenience, safety, reliability and cleanliness, he suggested that the invention was already well-established. For example, the lamps were referred to as 'burners', the bill for electricity was made for 'light hours' and the strength of the lamps was about the same as for a conventional gas jet.

A feature of the Edison distribution system was the use of a *parallel* method of lamp connection whereby *each lamp was placed directly across the mains lines*. The advantage of the method was that each lamp drew its own current. If one of the lamps became defective it would not prevent the rest working. Filament lamps operated in this way were rated at about 100 volt. This method of connection was a great step forward, for with arc lamps, which operated at a very much lower voltage, where several were in use with a single generator, they were connected in *series* so that the *total current flowed through each one*; the total generator voltage was set according to how many were in series, usually 2 or 3 arc lamps per generator, and if a fault developed in any one of the lamps, the others had to be taken out of service.

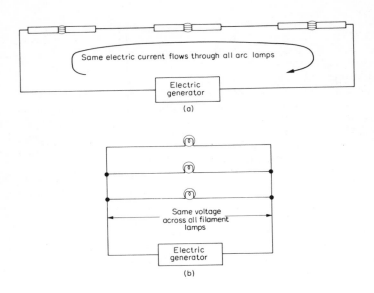

(a) Series lamp connections
(b) Parallel lamp connections

Electric Lighting Acts

By 1878 a large number of applications had been made by private companies
to install public electricity supplies in particular areas. They were in the same
position as the gas and water companies a few years earlier. A Select Government Committee was set up to examine the situation and make recommendations. It was generally agreed that electricity would be extremely important
in the near future and the basic problem was who should be in charge of the
undertakings. Should they be operated by private companies or by the Local
Authority? In other words which would be the best for the infant electrical
industry — free enterprise or a degree of nationalization? At this time there
was a distrust of large monopolies and implicit faith in competition between
private companies. The successful companies gave the best service at the
cheapest prices, unsuccessful companies becoming bankrupt and being
replaced by more efficient ones; therefore the public good was best served
by private enterprise. However, it was quite ridiculous to expect a number
of different electricity companies to supply electricity in the same street,
since this would have meant the duplication and even the triplication of
cables. If a customer decided to change companies the street would have to
be dug up again to make a reconnection, resulting in chaos for all and
exorbitant expense. A compromise was reached in the Act of 1882.

Brighton Power Station in 1887. This station is generally regarded
as the first viable public supply station. The plant installed in 1887
consisted of two Compound Fowler engines of 150 kW driving
Brush direct current generators. The engines were 'semi-portable.'
(By courtesy of I.E.E.)

The Electric Lighting Act of that year gave the right to private companies
to obtain Provisional Orders to operate in particular areas. These Orders gave
the companies the right to erect overhead electricity lines or to dig up streets
to lay cables. The Act was bedevilled by one of its clauses, which stated that
if the Local Authority were not satisfied with the progress of the company
it could cancel the authorization at the end of twenty-one years. After the
passage of the Act, applications for Provisional Orders continued to pour in
but nothing was accomplished because the twenty-one year clause was too
severe. Private individuals were not willing to invest in companies which could
be compulsorily dissolved after such a short time.

The Act was amended in 1888 by extending the twenty-one years pro-
bationary period to forty-two years and this immediately resulted in a wave
of investment which stimulated the delayed growth in the industry. From
that date onwards the progress of electrification was phenomenal.

The Direct Current Era

One of the most famous of the early d.c. generating stations and distribution
systems was at Holborn, London. It not only supplied the local public incandescent lighting scheme but also private individuals who wished to receive
current. The power plant was a 90 kilowatt steam engine running at 350
revolutions per minute directly connected to an Edison 'Jumbo' generator.
The generator output was about 60 kilowatt, which was sufficient to supply
1000 of the early 100 volt carbon filament lamps, each giving the same light
output as a standard gas lamp. It is an interesting aside at this point to ask
on what basis such a voltage was chosen. Filament lamps were originally
made to operate at 100 volt because such a pressure difference gave a

reasonable light output and length of service; the light output was about the same as the conventional gas lamp. Further, and perhaps more important, 100 was a 'good round number' and therefore useful in calculations. Later it was discovered that a voltage drop always occurred between the generator and the consumer due to the resistance of the intervening supply cables, so it was necessary to have some pressure 'in hand'. To achieve this the generators were rated at 110 volt.

By 1892 at the station in Kensington designed by Crompton, there were seven machines having a total capacity of 550 kilowatt. This was also a direct current system but gave a choice of two voltages — one for lighting and a higher one for industrial applications. A large standby battery was also available for peak demand periods and at night when the generators were switched off. This arrangement was possible because the load was mainly for lighting. Power demand was so low by 11 p.m. and throughout the night that all the current required could be supplied by the 600 ampere-hour battery. It is good economics always to run generators at their full capacities, because at low output frictional losses absorb a large fraction of the input power. Therefore if these machines can be shut down and the small amount of current required supplied by batteries, a great deal is saved on the fuel bill. (One must not forget, however, that the batteries must be first purchased and conscientiously maintained afterwards.) The nightly shutdown enabled the machines to be oiled and maintained in safety.

The Kensington Court Company, as it originally was called, had an extensive and ingenious distribution network to transmit the power from the generators to the consumers. The supply cables were laid in tunnels originally designed for water mains, sufficiently large for a man to stand to his full height. The use of subways avoided digging the streets, allowed easy installation, inspection and maintenance, and the possibility of laying extra cables in the same ducts when the system was extended. The flat rate charge for electricity was about 2p per unit with rebates up to 20 per cent if the total consumption was very large.

Both the Holborn Viaduct and the Kensington Court Systems epitomized future development. It was more economical to generate power in one central station and provide an extensive distribution network than for each consumer to generate his own electricity, for his own needs.

Technical Developments

To account for these achievements it is essential to retrace a few stages to the time of Pixii and the ensuing technical stalemate of bigger and bigger magneto-electric generators. As early as 1825 William Sturgeon had shown that electromagnets powered by batteries could be used instead of permanent magnets. In 1855 a Danish engineer named Søren Hjørth obtained a British patent which, although never exploited, showed how the permanent magnets of a magneto-electric generator could be replaced by electromagnets powered

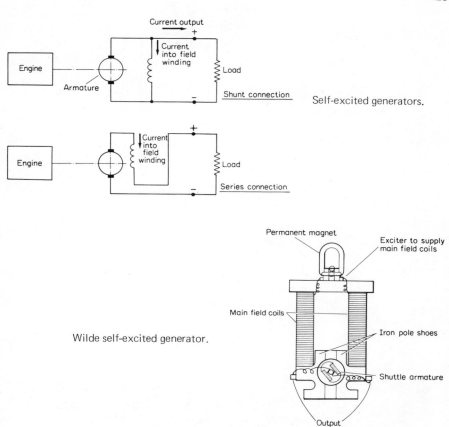

Self-excited generators.

Wilde self-excited generator.

from the generator itself. The initial exciting current was provided by batteries. Once the machine was running normally the currents for the electro-magnets could be supplied by diverting part of the electricity generated. Generators with electromagnets were called 'dynamo electric' machines and those which provided their own electromagnet current were said to be 'self excited'. Other inventors, such as S. A. Varley, C. Wheatstone, H. Wilde and the Siemens brothers, came to the same conclusions about ten years later and actually constructed working models. The basic design problem was how to start output from the electrical generator without using batteries; in other words, how to build up the magnetic field in the electromagnets initially to enable the generator current to begin.

Soon it was realized that the residual magnetism left in the soft-iron cores was sufficient to provide the necessary magnetic field to start generator

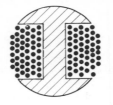

(a)

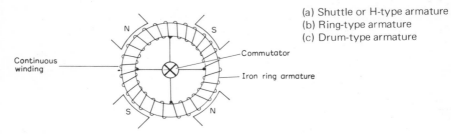

(a) Shuttle or H-type armature
(b) Ring-type armature
(c) Drum-type armature

Continuous winding

N   S

Commutator

Iron ring armature

S   N

(b)

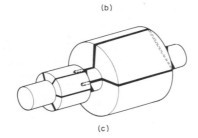

(c)

output. If this small initial current — or a fraction of it — were diverted through the electromagnets, thereby producing electromagnetism to augment the residual magnetism, then the total magnetic field would be greater and this would provide an even greater generator electrical output. With the continued working of the machine, both generator current output and field magnetism would build up naturally to values determined by the iron saturation and field winding resistance. By 1866 the principle of 'self-excitation' for direct current generators was established.

Alternating current generators or 'alternators' could also have electromagnets instead of permanent magnets, but in this case the field coils could not be excited by the alternator itself since it was producing alternating current and the electromagnets required direct current. A separate direct current supply from batteries or another direct current generator was necessary. (In fact the famous Wilde patent of 1863 on self-excitation related to an alternator and auxiliary direct current magneto-electric generator.)

Werner von Siemens, founder of the famous German electrical engineering firm, was one of the first to realize the importance of magnetic materials and circuits in generator design. He developed and patented the shuttle or H-type armature in 1856. This ensured not only that the coils were in a magnetic field of high intensity, but provided an excellent mechanical construction giving the possibility of high operating speeds.

The new design was intended to be used in a hand-driven magneto-electric generator employed in electric telegraph systems. Unfortunately one of its disadvantages was that when the armature and its conductors revolved, current

Early Siemens' dynamo (opposite). This machine is an example of the original commercial form of a drum armature machines. It was installed in the electric lighting plant of the Science Museum, London, in 1878 and ran until 1899. The drum armature was patented in 1873 by Werner von Siemens and F. von Hefner Alteneck. It consisted of a closed series of surface-wound coils provided with connections at regular intervals to a commutator of many segments, thereby delivering a fairly regular current. The armature consists of a hollow wooden drum, wound circumferentially with soft iron wire and provided with wooden pegs by which a longitudinal winding of insulated copper wire is held in position by brass bands. The field windings are in four elongated coils arranged to give consequent poles so that the magnetic field lines tend to run vertically through the armature. The machine has been partially sectioned but in its original state has the armature and field windings in series; its output was about 20 ampere at 50 volt.

(By courtesy of Science Museum.)

was induced in the armature core as well as the conductors. These unwanted circulating or eddy currents caused the generator to get very hot. In large machines water-cooling arrangements proved to be the only way to keep the generator working.

An important development in 1860 was due to Dr. Pacinotti, who suggested an alternative form of armature known as the *ring armature*, whereby the electrical conductors forming the complete turns cut the magnetic field lines in a more effective way. A number of individual coils were wound on a toothed iron ring and connected in series to form a continuous circuit. The connections from the joints between the coils were brought out to a multi-element commutator. When the ring revolved between the poles of a large electromagnet, the conductors on the outside of the ring moved so as to cut the flux lines at right angles. This is the most efficient way to generate a current. At the same time the ring-type armature design reduced the effect of the pulsating current in a highly inductive circuit.

Unfortunately Pacinotti's invention was not used until it was rediscovered by Zénobie Théophile Gramme some ten years later. Gramme's new machine was an improvement on Pacinotti's, for he employed a thicker ring and more coils which gave an even greater and smoother current output. The ring or armature core was sectionalized to reduce eddy currents. This was achieved by constructing the core of a continuous coil of soft iron wire, each turn being insulated from its neighbour by a thin coating of bitumen insulation. Within a short time Gramme's dynamos were being manufactured and sold on a large scale — he had a good business sense! These machines, which produced reliable if not large outputs, opened up a new era in electrical engineering and were used for arc lighting and electroplating. They could be run for indefinite periods without overheating.

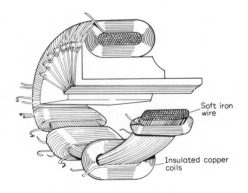

Gramme ring-type armature, 1870, cut to show the method of winding coils over the concentric wire core.

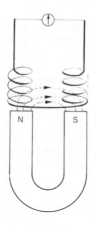

(a)

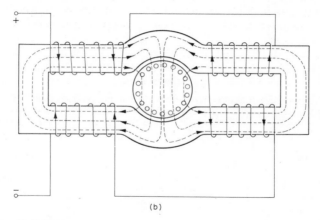

(b)

(a) Pixii Generator

This machine was a simple horseshoe magnet rotating beneath a pair of stationary coils. It was not an efficient electric generator because only a small fraction of the magnetic field cut the coil conductors at right angles.

(b) Hefner-Alteneck Generator.

This generator used the drum-type armature and magnetic field arrangement that ensured almost all the armature conductors cut the field lines at right angles.

Now that dynamos were being manufactured on a commercial scale, many inventors devised means of improving output, reducing losses, and bringing down costs. The operation of these machines was by this time basically understood.

A final major design advance was due to F. von Hefner Alteneck, chief
engineer of the Siemens-Halske company in Germany, who introduced the
drum armature in 1872. One of the disadvantages of the former ring armature
was that the inner conductors did not cut any magnetic field lines and therefore generated no useful current, serving merely to join useful sections of
outer conductors. The drum armature eliminated the need for centre conductors and had all the conductors on the outer surface of a solid armature: with
this arrangement all the conductors generate useful current, except, of course,
the end and commutator junction connections. Further, the drum armature
had another advantage over the Gramme ring in that it was possible to use a
preformed coil and no threading of wire was required. The first generators
employed conductors fixed to a solid wooden drum, but soon the construction changed to using iron. In the best designs, the coils were sunk into
longitudinal slots and held in place with wires or wedges. Shaped pole pieces
were fitted to the electromagnets to give a greater, more uniform spread of
magnetic field across the air gap. With these latter features it was possible to
reduce the clearance between fixed and moving parts, and this meant a
stronger magnetic field for a much reduced field current. In spite of the advantages of the drum armature, the Gramme ring continued in use for many
years.

Table 7
Comparison of Magneto-electric and Dynamo-electric Generators

	Type	Speed (rev/min)	Output (kW)	Weight (kg)	Cost £
Magneto-electric					
Holmes	Alternating current	400	2	2600	550
Alliance	Alternating current	400	2.3	1820	494
Dynamo-electric					
Gramme 1873 model	Direct current	420	3.2	1270	320
Siemens 1873 model	Direct current	480	5.5	930	265

Materials suitable for the new machines were not easily found. In the early
days insulated wires were only obtained by winding them by hand with silk
or cotton. The expansion of the electric telegraph industry in the 1850s led
to the investigation of ceramics for insulators. Wires insulated with
Guttapercha, india rubber and paper were gradually perfected towards the
end of the century.

These technical developments not only improved the performance of
direct current generators, but also benefited the direct current motor and
ultimately, of course, alternating current machines.

Motor Applications

Up to about the late 1870s the emphasis in the utilization of electricity had been in the field of lighting and electroplating, but application to motors was now becoming possible. The electric motor had been used in an experimental capacity in France a few years earlier for powering agricultural ploughing machines. The first notable application, however, was due to Edison, who devised a miniature pen motor powered by a twin-cell battery and running at 4000 rev/min for use in producing punctured copying stencils. After wide advertizing in the Edison style, it was selling at 60 000 a year. This is a startling example of miniaturization and mass production years ahead of its time.

In 1879 Siemens and Halske demonstrated the first non-battery electric railway in the world at the Berlin Industrial Exhibition where a demonstration line was laid down some 300 metres long. Electric current was picked up from a central rail insulated from earth working at 125 volt and returned via the earthed running rails. The motor was about 2 kilowatt and capable of a top speed of 8 kilometre/hour. During the short exhibition period it carried over 100 000 passengers. The insulation of the central rail left much to be desired, as it would only work on dry days. Two years later the first city electric tramway was installed at Lichterfelde, near Berlin. Electric power was fed to the cars by means of the running rails, but in wet conditions short circuits occurred between them and horses and pedestrians attempting to cross the track received shocks! Confidence in the electric motor was strong, however, and rail and tramways became important customers. It was indeed fortunate that the speed/power characteristics of direct current motors were so suitable for traction purposes that this application was pursued.

In 1883 the first British electric railway was built by Magnus Volk on a stretch of beach at Brighton. The system was similar to the German design, employing a central power rail at a similar voltage. Not to be outdone with such a tourist attraction, Blackpool built its own tramway in 1885 using a 'safer' conduit system. The third rail was suspended in a slot between the running rails and electrical contact was established to the vehicle by a sort of 'plough'. Both the central rail and conduit systems were later almost universally replaced by current collection from overhead wires. Leo Daft from New Jersey, U.S.A. was a most notable inventor of such systems and devised a small carriage or 'troller' which used to run on the wires to collect the power.

A twelve-mile street railway was installed in Richmond, Virginia, U.S.A. in 1887 and the contract at Richmond was won by the Sprague Electric Railway and Motor Company. Its success was due to the fact that their motors, which were carried on the tram axles, were very well designed and completely enclosed to prevent the entry of dirt and moisture. Since Sprague had to provide a system that would run thirty cars simultaneously, the current required was very large indeed. Individual cars carried a pantograph to collect the power from overhead wires. At first many residents bitterly complained of the ugliness of all the wires but these were forgotten — although not entirely —

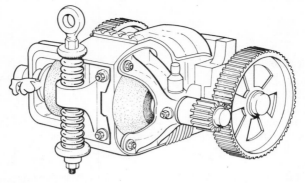

Sprague's traction motor (above). Each motor was mounted on an axle of the tram and had therefore to resist the entry of dirt and moisture as well as being small in size.

London Transport Electric Tram (below). From 1903 onwards the London County Council electrified most of their tramways as well as extending them, and by 1915 the last horse-drawn tram was withdrawn from service.

The tram depicted is one of the HR2 class built in 1930–31 for use on hilly routes — hence 'HR'. This class of car had two equal-wheeled trucks and four 26 kilowatt motors. Some of the HR2 class — including number 155 — were not equipped with trolley poles and thus were confined to the conduit part of the system. Route 11, with its long haul up the famous Highgate hill to the village, was one of the main preserves of the HR2 class. The last tram in London ran in 1952.

(By courtesy of London Transport.)

when the advantages of the new transportation were realized. Soon the horse-drawn tramcar was off the streets. One electric motor could do the work of two horses.

The electric trolley bus eventually replaced the electric tram, partly because of the high cost and difficulties of track mainten-ance. A typical example is afforded by Newcastle-upon-Tyne, where trolley buses were introduced in 1935. Their peak use occurred in the early 1960s, but by 1966 they were completely superseded by petrol or diesel omnibuses. The photograph, taken in 1948, shows a standard three-axle vehicle on one of the main routes from the distant suburbs to the city centre.
(By courtesy of Tyneside Passenger Transport Executive.)

The Leeds Corporation tramway employing overhead wires dates from 1891 and it set the pattern for development in this country. Each urban area set up its own enterprise and at one time there were 200 city tramways in regular use. As a typical example, West Hartlepool had steam trams from 1884 and electric trams in 1896. The service started with nine double deck, open-top, four wheel cars, and when the Corporation bought the tramway from the British Electric Traction Co. in 1912 there were eleven route kilometers operated by thirty-two trams.

The advent of tramways had important social repercussions. Cheaper transport was available to the middle classes who were then able to live in more pleasant surroundings and commute each day to their work in the city. In other words, improvement in transport was concurrent with the rapid outward expansion of our cities.

Tramways have virtually disappeared in Britain. (They are still popular, however, in continental cities.) The tramway was replaced by the trolley bus which gave a smoother ride, could pick up at the kerb and work through traffic, and then by the petrol or diesel bus which are the ultimate in cheapness and flexibility of routing.

One advantage of the electric motor over the internal combustion engine is that it produces, no atmospheric pollution. This superior feature was demonstrated by their use on the London Underground or tube system. The City and South London Tube was opened in 1891. The same voltage (550 volts) was used as on the tramways but the current was collected from a third rail instead of from overhead lines. The early London Underground system employing steam trains was really a system of lines sunk in 'cuttings' — if the line had to pass under roads, buildings etc. it could only do so for a short distance. A true underground railway is only possible with electric traction.

The London, Brighton and South Coast electrified railway was inaugurated in 1909 at 6600 volt alternating current, but that is another story.

An electric locomotive built in 1890 for the City and South London Railway (below, left) — the first major electrically operated railway in Great Britain and the first tube railway in the world. Heavily loaded trains were hauled at an average speed of 20 kilometre/hour, the maximum speed being 32—40 kilometre/hour. The locomotive runs on four wheels with a wheel base of about 2 metre the frames being outside the wheels. Each axle is driven by an independent electric motor which has its armature built directly on the axle. The field magnets embrace the armatures and are supported by bearings on the axle and by links from the locomotive frame. The Gramme ring-type armatures are series connected to their respective field coils. The electrical conductor is a steel channel placed between the rails and carried on glass insulators. Each of the two motors is rated at about 38 kilowatt.
(By courtesy of Science Museum.)

This Crompton electric traction motor is one of a pair installed in locomotive No. 36 of the City and South London Railway (below, right). The motor was in service for twenty-three years from 1900—23. Notice the motor drive directly on to the axle of the locomotive and the robust construction of the motor.
(By courtesy of Science Museum)

7 Electricity-the Final Growing Pains

In all this fragmentary development there was little attempt at organization or standardization on a large scale. Each installation, comprising generator, wiring, and control gear, was made with a specific application in mind, and various alternating and direct current systems were used. The competition between the two methods was called 'the Battle of the Systems'. It is well illustrated in the field of arc lighting where the two main firms involved were the Brush Company of the United States, who marketed a direct current lamp and the Société Générale d'Electricité of France, who marketed the Jablochkoff alternating current lamp.

In an arc lamp, an electric arc is formed between the free ends of two carbon rods which become incandescent and burn away. Both direct current and alternating current can be used to power the lamps, and while the former gives the greater light output, it also causes the rods to be consumed at different rates. The Brush Company solved this problem with a rather complicated slipping clutch mechanism for one of the rods which was operated by current flowing in the arc. The Jablochkoff alternating current lamp was far simpler. It consisted of two carbon rods mounted vertically which acted like a candle; the discharge was started at one end of the rods and moved slowly down their length, both rods being consumed at the same rate.

With the introduction of the Edison-Swan carbon filament lamps, which could operate equally well on alternating or direct current supplies, it seemed that the struggle for the 'best' system as distinct from the best lamp was beginning in earnest.

Superiority of Alternating Current

Near the end of the nineteenth century the pattern of daily electricity demand from the direct current central stations changed. With the new tramways, railways and factory processes the demand became larger and more continuous. Extra machines were installed at the generating stations but the standby batteries became an anachronism and were usually converted to power supplies for energizing switches, instruments, and relays, and for providing a station emergency lighting supply.

133

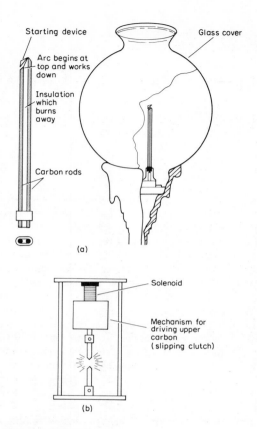

(a) Jablochkoff Candle.
(b) Brush Arc lamp. Note the control mechanism is arranged
  so as not to throw a shadow.

The future of direct current took a further turn for the worse with the invention of the alternating current induction motor, which is basically simpler in construction and more reliable than an equivalent direct current motor, because it has no commutator requiring regular servicing due to constant sparking and brush-wear. The induction motor, however, is essentially a constant speed machine, control of the speed being difficult by comparison with the direct current motor. Fortunately for the alternating current motor, most power drives in factories required a constant speed. The motor was used to drive long line shafts running the length of workshops, and pulleys mounted on the line shafts transmitted power to individual machines by thick belts. One of the first factories to be electrified was a cotton mill in the United States, which went into operation in 1894. The mill was a

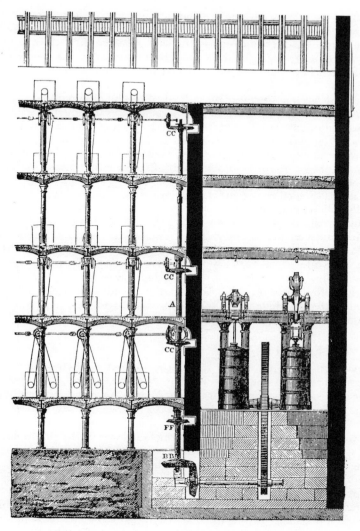

Section of a Woollen Mill about 1850, showing the method
of driving shafts, gears and belts necessitated the production
machinery being installed as close as possible to the driving
engine, in this case a beam engine of 75 kW.

(By courtesy of Mills and Millwork by Sir W. Fairbairn.)

completely alternating current installation; instead of a central steam engine, there was a central alternating current motor which drove the line shafts. At this time firms like Westinghouse in the United States and A.E.G. in Germany were important producers of these induction motors and there were many bitter quarrels between them over infringement of patent rights.

It was not until well into the twentieth century that motors were attached to individual machines. Nowadays individual machines may have many different motors operating the various parts. Electric conductors carrying the power run the length of the workshops, replacing the old line-shaft, and supplies are tapped off for the machines at the most convenient points. Further, the flexibility in the supply of electricity enables the greatest freedom of choice not only in the design of factories but also in the siting of factories.

The alternating current generator had many technical advantages over an equivalent direct current generator, the chief amongst these being that, like an induction motor, it needed no commutator. More important still was the ease with which an alternating voltage could be changed in magnitude with the aid of a transformer, using the principle discovered by Faraday fifty years before. The voltage ratio between the input and output sides of the transformer was determined solely by the turns ratio, and it was not necessary to make generation and utilization voltages equal. Electrical power could be produced and transmitted at a high and economical voltage and then stepped down to a lower range more suitable for consumption.

Sebastian de Ferranti

The first engineer to put the above ideas into effect was Sebastian de Ferranti, who worked for the London Electric Supply Corporation. At an early age he became a passionate supporter of the alternating current system and turned his creative genius to furthering its cause. In 1882 he invented a new type of alternator with a special zig-zag armature which gave an extremely high output power for its size. After becoming chief engineer of the Grosvenor Gallery power station four years later, he completely reorganized it within a few months so that it worked at the much higher voltage of 2400 volt. It is especially interesting that he connected his transformers in parallel across the supply, instead of in series as had been usual up to that time.

Ferranti was responsible for designing the first great power station in the world and indeed may be described as the 'father of electricity supply'. He saw the growing need for electric power in the city and realized it must be met on a large scale with a really big generating plant. A well-planned power station needs space for its various facilities, and since land was expensive near London the station had to be built some distance away. A good site was found on the Thames at Deptford, some twelve kilometres from the City. This was rather a large distance for those days and the power had to be transmitted at a high voltage for the scheme to be economically viable. A 10 000 kilovolt transmission voltage was chosen which was far higher than anything else

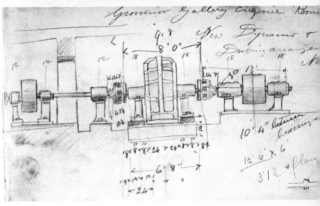

Ferranti 550 kW alternators at the Grosvenor Gallery 1887
(above, left).
(By courtesy of Science Museum.)

Page from Ferranti's sketch book showing design for alternators
at the Grosvenor Gallery (above, right).
(By courtesy of Science Museum)

previously attempted on such a scale. Needless to say the problems
encountered were enormous. No long lengths of cable to operate at that
voltage had yet been made so Ferranti had to set about developing them. (The
high voltage cables he originally ordered were not satisfactory even at only
half the voltage specification.) Fortunately he was able to obtain permission
to run his electricity into the City alongside the railway lines both above and
below ground and across bridges. Twelve kilometres of road digging would
have been a major item in the cost of the scheme at that time. Ferranti's
cables used rigid copper tubes insulated with waxed paper; a substantial
earthed metal case provided mechanical protection. An amusing story is told
of the Board of Trade officials who insisted on a full test of the new cables.
To prove they were safe, one of Ferranti's engineers arranged a demonstration
in which he chiselled through the live cable. Fortunately the earthed outer

Swan carbon filament lamps. The experiment lamp on the left is one of those made by Sir Joseph Wilson Swan in 1878 and exhibited in Newcastle in February in 1879. The filament was of carbonized paper, held between platinum wires and kept burning during the final stages of evacuation to remove occluded gases.

The commercial lamp on the right was made about 1880. This had an improved filament obtained by carbonizing specially treated cotton thread.

Three years later he developed a superior filament made with artificial silk.

(By courtesy of Science Museum.)

cover through which the chisel penetrated performed its function satisfactorily, the live conductor was immediately earthed and the engineer lived to tell the story.

It was a commonly-held belief at this time that electric shocks received from alternating current were more dangerous than those from direct current. Like most 'words of wisdom' there is usually a grain of truth somewhere. What the objectors to alternating current were really saying was that it was more difficult to estimate insulation for alternating current cables than for direct current cables. This is because alternating voltages are specified by the root-mean-square or effective value rather than the maximum value, whereas it is the latter which must be considered when calculating the necessary insulation.

The distrust of alternating current is epitomized by remarks made by Edison in one of his 1887 advertizing pamphlets. 'Take Warning! Alternating currents are dangerous! They are fit only for powering the electric chair. The only similarity between an alternating current and a direct current lighting system is that they start from the same pile of coal.' Adding further to this distrust, it was found that when a cable was connected to an alternating current supply a number of unanticipated effects occurred. For example, the voltage at the receiving end was greater than that at the sending end; only

Map of overhead distribution system from Grosvenor Gallery
Station, London 1887 — later connected to Deptford power
station by a 10 000 volt transmission line.
(By courtesy of Science Museum.)

8.5 kilovolt at Deptford was sufficient to produce the required 10 kilovolt in London. This rise of voltage along the line became known as the 'Ferranti Effect'. It was eventually explained as resulting from an interaction between the cable itself and the transformer supplying it; the cable possessed a kind of electrical resilience or capacitance and the transformer an electrical inertia or self inductance. Both these effects were only familiar to direct current engineers when switching the current on or off but they were always present with alternating current.

Parallel-Operation and the Turbo-Alternator

On the generation side one of the greatest problems to overcome was that of connecting alternators in parallel. Direct current generators were easily joined together electrically to function as one unit if their voltages were 'equal'; however with alternating current machines there were added difficulties because the machines had to give exactly the same frequency, and the magnitude of the phase voltages had to be equal. Large-size generators powered by slow-speed reciprocating steam engines were difficult to match in speed and phase, and even worse, the speed flucuated quite markedly during each cycle. No wonder alternators were difficult to synchronize! In bringing the two machines to work together the engineer-in-charge stood a good chance of burning out all the lamps connected to the supply and of permanently damaging the generators. The problem was only solved with the invention of the high-speed turbo-alternator which had a more uniform speed during each cycle. It allows synchronizing to be done with a much better chance of success.

The steam turbine was invented by C. A. Parsons in 1884 and was hailed as the greatest advance in prime movers since Watt's steam engine. The principle of directing a steam jet on to the blades of a fan to cause motion had been known for years but Parsons developed the idea into a real machine, a practical turbine. By paying attention to such detail as the shape of the turbine blades and their best arrangement along the turbine length, he improved its performance beyond recognition. The turbine works best at high speed, and his first designs worked at the phenomenal speed of 18 000 rev/min and were coupled to d.c. generators used to supply current to lighting installations on ships. In 1888 he produced bigger turbines working at the lower speed of 4800 rev/min and these were used to drive 75 kW, 1 kV alternators. They were the first turbo-alternators.

By the turn of the century Parsons' success seemed unlimited. In 1897 he fitted his revolutionary turbine to the steam yacht 'Turbinia' and caused a sensation at the Spithead Naval Review by steaming between the lines of warships at 35 knots. Soon his turbines were fitted in the big passenger ships 'Mauretania' and 'Lusitania', as well as the new Dreadnought class of battleships. In 1899 he built two 1000 kilowatt, 4 kilovolt, single-phase turbo-alternators, with a running speed of 1500 revolutions per minute, for

The London Electric Supply Corporation Limited,

3, ADELPHI TERRACE, LONDON, W.C.

ELECTRIC LIGHTING.

HIGH PRESSURE SYSTEM v. LOW PRESSURE SYSTEM.

The Corporation supplies Electrical Energy on the High Pressure System by Meter.

THIS SYSTEM, when properly arranged and controlled, as it is by this Corporation, is **equally as safe as the Low pressure system**, besides which it has the great advantage of giving an

EQUAL AND REGULAR LIGHT

throughout the whole of the premises, however distant the farthest lamp may be from the point of supply.

An idea appears to exist that the High pressure system is more dangerous than the Low pressure system. This is erroneous. The following is the opinion of Mr. W. H. PREECE, F.R.S., the Engineer-in-Chief and Electrician to the General Post Office, as expressed by him in his address as President for the year, at the Meeting of the Institution of Electrical Engineers, held on Thursday, the 26th January, 1893.

"**The prejudice against High pressure is still strong, it is thought to "be unsafe, but time and experience will eradicate this impression as they "ultimately eradicate every fallacy.**"

The following is a diagram of the system and connections.

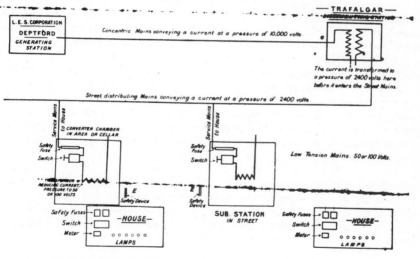

It will be seen that the High pressure at the point of entering the Consumer's premises is **Reduced to a pressure of 50 or 100 volts**, which is similar to the pressure of the Low or Continuous Current system.

At no point are the High pressure supply wires in contact with the Low pressure service wires.

This Corporation supplies current to H.R.H. THE PRINCE OF WALES, at Marlborough House, a large number of the Nobility, Clubs, Hotels, Theatres, &c., &c., **where a good and continuous light is the first consideration.**

SPECIAL TERMS ARE QUOTED TO LARGE CONSUMERS AND TO THOSE USING THE LIGHT FOR UNUSUALLY LONG HOURS.

[P.T.O.

Leaflet issued by London Electric Supply Corporation Limited 1893.

(By courtesy of Science Museum.)

Elberfeld in Germany. They were the biggest generating sets ever built and had the highest-ever steam efficiency. Manufacturers of electrical power plant all over the world were very impressed and electrical supply companies were delighted that such performances could be achieved. Success bred success. Parsons' factory was unable to meet the flood of orders and the manufacturing rights were sold to others.

However, Parsons did not concentrate solely on the mechanical aspects of electrical power production. He devoted much effort to improving the alternator itself and tackled the electric, magnetic, and ventilating circuits of the generator in turn. Detailed investigations of such problems as conductor insulation, eddy currents in the iron, and the use of ventilating ducts to aid heat removal were successfully pursued. In the course of this machine development Parsons switched in 1902 from a rotating conductor system to a rotating electromagnet system, in which the revolving magnets had projecting poles. This system was adopted for the 1500 kilowatt, 6.6 kilovolt alternators for the Neptune Bank station of the Newcastle upon Tyne Electric Supply Co. The occasion marked an important advance in the construction of high speed alternators, and the opening of the station was performed by Lord Kelvin.

Alternators could have 2, 4, 6, 8, or more projecting poles. In general the greater the number of poles, the greater was the machine diameter and the lower the speed. Since turbines worked best at high speeds, most alternators employed 2 or 4 poles and were of a relatively small diameter.

A major design problem of high speed alternators was to make the rotating member strong enough to resist the tremendous bursting forces caused by rapid rotation but at the same time it had to be perfectly balanced in order to avoid dangerous vibrations. Brown-Boveri, a Swiss company, suggested a partial solution. If the rotating member were made from solid steel it would be immensely strong and the conductors could be embedded in deep slots cut in the steel. In 1911 Parsons manufactured a 300 kilowatt, 4000 revolutions per minute alternator using the solid forging technique.

Two years before World War I a turbo-alternator of 25 000 kilowatt was installed in a Chicago power station. To a great extent this tremendous growth in capacity was obtained by the use of artifical cooling. The best method discovered was to enclose the whole generator in order to control the flow of cooling air, and this had the added advantage of noise reduction. In a short space of time the world had become covered with a complex of power stations.

Electricity Supply Companies

A typical pioneer supply-undertaking was the Newcastle and District Electric Lighting Co. in Parsons' own area. The Company was formed in 1889 with the Forth Banks station and laid down electricity mains in the western half of the city. Single-phase alternating current was generated at 80 hertz. Both generation and supply mains voltage were 1 kilovolt, with step-down transformers to 100 volt at the consumers' premises for lighting. In the first instance there

Table 8

Growth of Newcastle-upon-Tyne Electric Supply Co.

Generating Station	Commenced Supply	Total Output Capacity (kilowatt)	Steam Pressure (bars)	Steam Temperature (°C)
Pandon Dene	1889	2400	8	180
Neptune Bank	1900	4300	14	200
Carville 'A'	1904	31 000	14	280
Dunston 'A'	1910	70 980	14	320
Carville 'B'	1916	55 000	19	360
North Tees 'A'	1921	40 000	33	360
North Tees 'B'	1923	70 000	33	360
Dunston 'B'	1933	150 000	44	440

were seven small generators to provide the current, but later the capacity was much increased with the installation of two 400 kilowatt and three 500 kilowatt machines.

Their second power station at The Close was commenced in 1902 to supplement and eventually replace the Forth Banks station. The first stage was completed in 1904, when it housed two 1000 kilowatt, 500 volt direct current generators. Direct current was urgently required for the rapidly expanding city tramways and new coast railway. An extensive system of direct current mains supply cables was laid for other consumers. Up to 1908 The Close station was extended in sections by the addition of further direct current generators and three alternators from Forth Banks. In 1915 and 1916 the generation of 40 hertz three-phase alternating current was begun.

Early turbo-generator of C. A. Parsons (1884). The double-flow turbine operated with steam at 5.5 atmospheres of pressure 160° C exhausting to atmospheric pressure giving a shaft speed of 18 000 revolutions per minute. A two-pole dynamo with an armature employing oil cooling gave an electrical output of 7.5 kilowatt (75 ampere at 100 volt).

(By courtesy of Reyrolle-Parsons Ltd.)

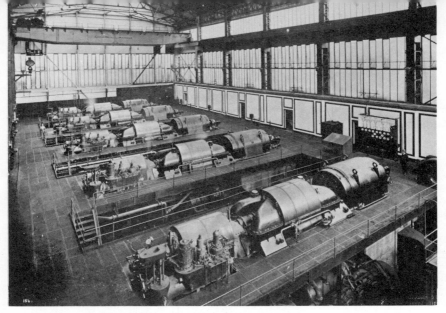

Carville 'B' power station, 1920, owned by the Newcastle and
District Supply Company (below). The plant comprised five
11 000 kilowatt (11 megawatt) 40 hertz turbo-generators running
at 2400 revolutions per minute. Machines of this size imposed a
number of problems in design. The generator was very long compared with existing practice and it was evident that a simple axial
air cooling system would be insufficient. Parsons decided on a
new system which provided a multiple-path radial and axial flow
system for the stator and water cooling of the rotor.
(By courtesy of Reyrolle-Parsons Ltd.)

With three electrical systems in use by this single company, it was evident
things were getting a little out of hand. To enable power transfers between
the systems, a 750 kilowatt rotary converter was installed to change the
alternating to direct current, and a 40/80 hertz frequency changer was
introduced to convert the power at one mains frequency to power at the
other mains frequency.

At the same time in the eastern half of the city, the Newcastle upon Tyne
Electric Supply Company was developing a completely separate network at
100 hertz and 40 hertz, the latter frequency being the more important of the
two in subsequent development.

The situation in Newcastle was characteristic of the general trend, but
eventually economic, social and strategic pressures forced the proliferation of
systems to cease. (The Newcastle upon Tyne Electric Supply Company
eventually took over the District Lighting Company as well as all other local
suppliers north and south of the river.) Alternating current won the day and
standardization of voltages and frequency were established, but not fully completed until 1960. This is the subject matter of the next chapter, which
describes how order developed out of chaos.

8 Electricity Comes of Age

Piecemeal Development

With the more widespread use of electricity in factories and houses demand began to reach very large proportions, and during World War I the shortcomings of the nation's supplies became apparent. The piecemeal development had many disadvantages and there was a great need for better organization and rationalization. For example, if a town's generators failed, the whole town would be completely without electricity until the defect was repaired. The ensuing confusion can be imagined. It was absurd that local plant failures should cause so much chaos, especially in the early days when it happened quite frequently. Neighbouring towns which could possibly have helped out in the emergency were unable to do so because there was no connection or transmission line linking them. Even if there had been it would be useless in most cases because different towns used different systems. In 1924 there were nearly two hundred direct current systems in addition to the alternating current systems which employed seventeen different frequencies. This parochialism also meant that each authority had to provide its own spare plant.

World War I also showed the need for more widespread electrification and the country had to be considered as a single area. Not only should the remaining urban areas be electrified but the rural areas too, even if the latter were unprofitable. Electrification enabled food production to be increased so that the nation could be more self-supporting, and this was vital for our security in times of war — submarine warfare had shown how vulnerable were food supply convoys from abroad.

The National Grid

Great political agitation and pressure occurred after the war, and much rethinking was done so that by 1926 the remedy was in hand. The Electricity Act of that year established the Central Electricity Board whose function was to erect a unified network of interconnecting power lines over the whole country and to standardize the frequency. This is the National Grid. The name is derived from the Weir Committee Report of 1927 which described the network of transmission lines as looking like a 'grid iron' pattern on the map.

The overhead transmission lines were fed by the important generators of each town or city and, ideally, each urban area was to be connected to the

145

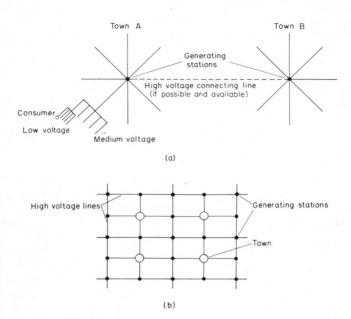

(a) Radial (pre-1926, above) and (b) Grid (post-1926, below)
Distribution Systems.

Grid matrix at a minimum of two points to give a better guaranteed supply.
Transformers and high voltage lines working at 132 000 volt enabled power
to be transmitted economically between regions, for use when one region
had a surplus and another had a deficiency of power. This was in contrast with
the previous arrangement in which each locality was utterly dependent on a
single supply line from one central power station. One of the largest single
benefits of the 'Grid' was its ability to reduce the *total* amount of standby
plants; each authority connected to the Grid did not need not to provide its
own full back-up equipment in case of breakdown, and relatively few spare
generators were sufficient to cover most eventualities. Former idle plant
could then be used to meet the expanding demand for power with consequent
economies in capital investment.

A prerequisite for Grid operation was that the big supply power stations
worked at the same alternating current frequency, since power at different
frequencies could not be fed on to the same Grid lines, standardization was
established at 50 hertz. Companies with non-standard generators were not
linked to the Grid and their generators were gradually phased out. It must be
remembered that a large amount of capital was involved in this decision as at

least eighty undertakings operated at frequencies other than 50 hertz. Financial compensation was given to supply companies, firms and private householders who 'suffered' as a result of standardization.

Where direct current supplies were essential, converters or rectifiers were installed to change the alternating to direct current; the simplest but unfortunately the most expensive method used a two-machine arrangement — an alternating current motor driving a direct current generator. Special rotary converters or banks of mercury arc rectifiers were later established where really big outputs were required. Factories and tramways usually housed this important plant in specially allocated substations.

On commercial operations the Central Electricity Board, i.e. the body established to build and operate the Grid, purchased electricity directly from the generating stations and sold it to authorized supply companies at cost price, the generating station companies acting as 'producers', the board as 'wholesalers' and the supply companies as 'retailers'. The charge to each retailer or distributor was the same, irrespective of region. Financially, therefore, some areas gained and others lost. Formerly, areas with high concentrations of industry and efficient generators, such as Tyneside, marketed cheap electricity, whilst other areas, especially rural districts, paid highly (In some cases the cost might have been as much as ten or fifteen times more.) With the new scheme the areas with the better class of generator subsidized the others. Whatever the rights or wrongs of the arrangement, most engineers agreed with the change. Electrical engineering pride in local plant was replaced by a more justified satisfaction in a good national system. The National Grid was one of the first attempts at nationalization, preceded only by the Forestry Commission which was set up in 1919; the B.B.C. followed, in 1927. Like these the Central Electricity Board had a constitution as a private industrial company, in that once the members of the Board were appointed they were free from direct day-to-day control either by a minister or by Parliament.

During the inter-war years, technical development raced ahead. A steam turbo-alternator of 105 megawatt was installed at Battersea, London, in 1933. More and superior turbine blades, better steam conditions and improved alternator cooling all helped to increase electrical output. The increased power-flow from generators to the National Grid led to parallel development of associated equipment. Both the transformers and the switches controlling current flow were improved. Manufacturers set up their own development and research laboratories to improve their products, for example the Reyrolle Short-Circuit Testing Station founded at Hebburn in 1929 and the Nelson Research Laboratories at Stafford just prior to World War II.

Nationalization

In 1936 a government-appointed committee under Lord McGowan issued a report on electricity supply. Its conclusion was that drastic reorganization

was needed, and it suggested this could be done either on a regional basis
under public control or by the compulsory absorption of the smaller undertakings into larger units. Legislative action on these proposals was frustrated
by the outbreak of World War II. However, there were very many critics
of this centralization. They argued it was foolish to put 'all one's eggs in
a few baskets' since- enemy action at these vulnerable points by aircraft
bombing or sabotage could easily cripple the nation's electricity supplies.
The arguments were in vain and the ensuing war did not halt this planning
for centralization. The existing Grid system, operating at 132 kilovolt, enabled
some industries to be set up in remote areas and other industries to be moved
from vulnerable areas.

During the war plans were laid not only for greater centralization but also
for the *complete* integration of the whole of the electricity supply industry.
Soon the long-term planners were thinking of a new, larger Grid (or Supergrid)
to operate at twice the existing voltage, i.e. 264 kilovolt. This was bold thinking indeed!

At the end of the war demand for electricity rapidly increased and it
became obvious that a major reconstruction programme was needed to
replace old inefficient plant and to cope with increasing requirements. A
series of bad winters coupled with fuel shortages and power cuts emphasized
the seriousness of the situation. The Labour Party, who were in office at this
time, regarded the profit motive and big business man with deep distrust and
believed that all key services and industries should be under government
control. The electricity supply in the country was particularly of national
importance and should be under public ownership. The complete nationalization of the supply industry was carried out in 1947. Vesting day was 1 April
1948 and from that date the whole of the business of the electricity supply
was taken over by the British Electricity Authority. Its duty was defined as
the development and maintenance of an efficient co-ordinated and economical system of electricity supply, and as such it was responsible for the generation and transmission of electricity in bulk and for supplying consumers via a
number of Area Boards. The Authority was really a giant commercial enterprise devoted to two major activities — (a) producing and wholesaling and
(b) retailing — the commercial equivalent being:

> production   — power stations
> wholesaling — the Grid system
> retailing   — the Area Board networks

General control was exercised over the Area Boards by the central authority.
The country, except the North of Scotland, was divided into fourteen generat-
ing divisions which coincided with fourteen Area Boards. (The North of
Scotland Hydro-Electric Board was inaugurated much earlier in 1943.) The
Area Boards purchased electricity from the central authority and distributed
it to consumers.

The political organization of the British Electricity Authority was very different from previous public enterprises. Ownership of the industry was transferred from directors and shareholders — with due compensation — to the Board or Corporation appointed by the Minister of Power, who was responsible to Parliament for the Authority.

There were many problems facing the new organization. Over five hundred separate bodies had been responsible for the generation and supply of electricity and there were wide variations in plant and efficiency. During the war the

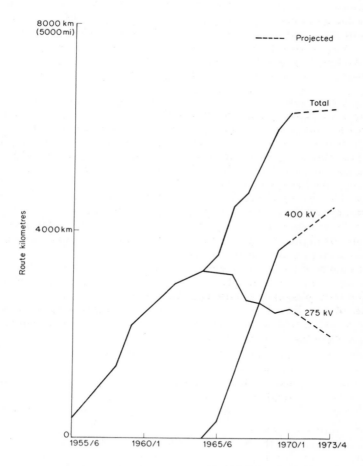

Route kilometres of overhead lines. 1955/56 — 1973/74.
The 132 kV transmission lines are excluded.
(By courtesy of C.E.G.B.)

building of power stations was greatly reduced so that there was need for
tremendous expansion, and this the Authority embarked upon immediately.
Between 1 April 1948 and March 1964, 78 new generating stations were
built, 19 new sections of existing stations opened and 47 large scale exten-
sions to other stations carried out. This policy of expansion raised the capacity
of the industry from 10 000 megawatt to 33 000 megawatt, a similar
expansion occurring on the high voltage transmission side. Not only was the
132 kilovolt system extended, but a new national Supergrid working at
275 kilovolt was begun in 1952. The 264 kilovolt level, which as mentioned
previously was first thought of during the war, was increased to 275 kilovolt
to conform with international thinking and to enable U.K. firms to compete
throughout the world based on their work on the National Grid. At the end of
March 1964 the industry owned almost 13 000 kilometres of overhead lines at
the two voltages in addition to several hundred kilometres of underground cable,
and, of course, the expansion still continues.

Development in distribution can be illustrated by the fact that annual sales
to consumers in 1947/48 rose from under 33 000 million units to more than
121 000 million units in 1963/64. This involved a considerable increase in the
number of consumers of almost every type — domestic, commercial, and
industrial. One remarkable increase was the number of farm consumers. On
Vesting Day only 31 per cent (85 000) of all farms in England and Wales were
connected to the mains, but by March 1964 the number had risen to 255 000
which equalled 91 per cent of the total.

In 1960 a study was made of the expected transmission situation in 1970
by which time it was predicted that the load would have increased two and a
half times to 77 000 megawatt. A new transmission voltage of 400 kilovolt
was adopted as standard since further extension of the 275 kilovolt system
would be complex and involve many more transmission lines than would be
required for the 400 kilovolt system.

The growth rate turned out to be an over-estimate because of national
economic difficulties encountered in the late 1960s.

In the 1970/71 period the maximum output capacity was 49 000 megawatt,
the predicted peak demand was 41 000 megawatt and the maximum demand
met was 39 000 megawatt. At the time of writing (1972) the growth of
electricity demand continues to be at a somewhat slower rate than in the past.
The predicted demand for the winter of 1975/76 is 54 000 megawatt.

According to politicians Britain can expect a much increased enonomic
growth rate after joining the European Common Market. If this is correct then
the load figure of 70 000 megawatt will probably be an accurate estimate for
the early 1980s.

Central Electricity Generating Board

Following a Government Committee of Inquiry a New Electricity Act was
passed in 1957, re-organizing the industry. The Act created the Central

Electricity Generating Board (C.E.G.B.) from the 1 January 1958 to control the generation and transmission sides of the business and The Electricity Council to co-ordinate the work of the industry as a whole. The Area Electricity Boards remained much as they were.

The C.E.G.B. control the operation and maintenance of power stations and supply electricity in bulk to the Area Boards in England and Wales via the Grid system. It is also responsible for the siting, design, and construction of new generating stations and transmission lines. The Board's organization is made up of a main headquarters, two Divisions (one responsible for Generation Development and Construction and the other for Transmission Development and Construction), and five Regions†, each a self-contained management unit. (The North of Scotland Hydro-Electric and the South of Scotland Electricity Boards are connected to the C.E.G.B. but administered separately.)

There are twelve Area Electricity Boards which buy bulk supplies of electricity from the C.E.G.B. They are responsible for local distribution networks and the sale of electricity to consumers in their Area. They also take part in contracting and the sale of electrical appliances in their own shops or service centres.

There are many technical advantages resulting from such an integrated approach to electricity supply. A few of these are worth mentioning and relate to research, planning, construction and operation.

1. Large organizations like the C.E.G.B. can finance major research projects. They may vary from novel ways of producing electricity, such as magnetohydrodynamic generation, to the investigation of 'dry' cooling towers.

2. There is no need to plan stations in relation to the area they are to serve; instead they are planned in relation to the source of energy for the prime mover. The high voltage Grid and Supergrid lines allow the power to be transmitted to the required bulk supply points efficiently and cheaply.

3. Power stations may be constructed without large amounts of standby plant. If a fault develops on one of the generators, extra power can be supplied from other stations on the Grid.

4. Very large and efficient power stations can be built and kept running as long as possible to produce the cheapest electricity. These are called 'base load' stations. Conversely, older and less efficient plant is run only at peak periods when maximum output from all generators is required.

The electricity Grid today may be regarded as the logical vindication of Ferranti's philosophy of large-scale generation, high voltage transmission and efficient distribution.

---

† For administration purposes, the two Midland Grid Control Areas are joined together, as are the two South Eastern Areas.

Central Electricity Research Laboratories

Research started in 1932 with investigations on the 132 kilovolt transmission lines of the then recently introduced National Grid, and these were the main studies until nationalization in 1948. In 1956 the Herbert Committee criticized the industry for relatively small financial expenditure on research, and with the re-organization in 1958 this was rectified. The research expenditure in 1957/58 was £0.76 million and in 1969/70 it was £9 million.

Today the C.E.G.B. has three large laboratories at Leatherhead in Surrey, Berkeley in Gloucestershire and Marchwood in Hampshire. The Leatherhead Laboratories carry out basic investigations into most aspects of conventional generation and transmission equipment, Marchwood Laboratories are mainly concerned with large-scale feasibility studies, and Berkeley Laboratories with the economic and safe generation of electricity in nuclear stations.

In addition, there are smaller laboratories in each of the five regions, so that research scientists and engineers are in close contact to the people who actually operate the generating plant and transmission system. A high proportion of regional research is directed towards solving particular, and sometimes local, problems.

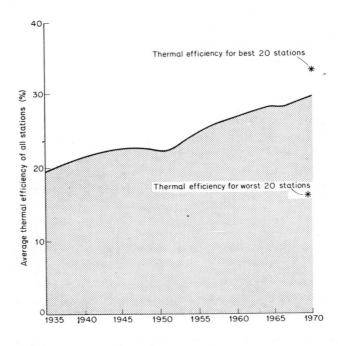

Thermal efficiency of power stations.

Free Enterprise

Perhaps the greatest responsibility for the phenomenal technical progress since World War II has been due to the manufacturers of electrical equipment. Most big companies and a large number of small firms have set up research and development groups. The International Research and Development Laboratories at Newcastle were founded by C. A. Parsons Ltd., to improve their products by a better understanding of the fundamental principles of their operation. The scheme was such a success that it became self-financing. Similar advanced laboratories were also developed by other companies. Technological development has improved the performance of generators, transformers and switchgear.

Generators are available with outputs up to 660 megawatt. Power transformers have efficiencies of 99.5 per cent, and high voltage switches can interrupt circuits carrying currents of nearly 100 000 amperes. More specifically, in the case of generators not only have outputs increased tremendously, but the electrical power obtained from a given amount of material has increased out of all proportion. As discussed earlier this has been achieved largely by the use of improved cooling systems. No doubt this trend to larger sizes and higher efficiency will be continued.

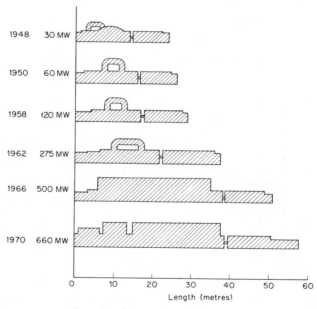

Size of turbo-generators.

Nuclear Power

Electrical engineers have always tended to adapt other technologies to the best advantage, the most notable example being the introduction of nuclear

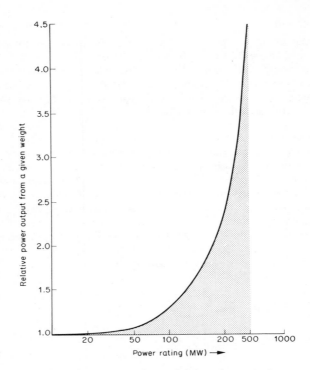

Increase of efficiency in use of materials in turbo-generator construction.

energy for the production of electricity. Calder Hall in Cumberland, opened by Her Majesty the Queen in October 1956, was the world's first full-scale nuclear power station to go into commercial operation. There are now in operation a large number of similar and more advanced types of nuclear reactor working, and up to 1970 British reactors had produced more electricity than the rest of the world put together. Nuclear reactors generate heat which produces steam to drive a conventional steam turbine and alternating current generator.

Industrial Disputes

During the winters of 1970/71 and 1971/72 industrial disputes caused the power supply industry to assume a new political force. In January and February 1972 the coal miners successfully 'picketed' the big power stations, preventing the movement of coal and using this as a means of bargaining in their wage negotiations. Their action caused voltage reductions and power cuts affecting the whole country. A pay settlement was eventually arrived at following a Special Inquiry and the miners obtained as a result of their actions against another branch of industry most of what they claimed. The picketing tactics were highly successful and there is now some concern that in future disputes other unions may adopt these techniques in pay disputes.

Institution of Electrical Engineers

No account of the development of electrical engineering in this and most other countries would be complete without some reference to the professional institutions. Just over one hundred years ago, in May 1871, the Society of Telegraph Engineers was formed which later grew into the Institute of Electrical Engineers based in London. In 1969 there was a total membership of 60 000 including 10 000 members living outside the United Kingdom. The Institution has played an important role in promoting the advancement of electrical and electronic science and engineering. It has provided and still provides an extensive range of meetings, conferences, residential schools and publications designed to enable members to keep pace with progress in electrical, electronic and related subjects. The Institution sets the standards of qualification for professional electrical and electronic engineers.

Looking back over the last thirty years, with perhaps just a little exaggeration, the enormity of the progress made by the electrical engineer in his technology can be more readily appreciated if it were supposed that the Wright brothers had decided to build a jet airliner as their next project and had succeeded!

9 Epilogue

The purpose of this book has been to give you some idea of what electrical engineering is about and what electrical engineers do.

Engineering is the application of the findings of science to satisfy the material needs of society. An engineer is concerned with making things and getting things done, and he uses knowledge that science gives to help him achieve these ends. In particular, an electrical engineer or electrical technologist uses knowledge of electricity to solve his problems — although he must draw on many other fields of learning as well.

The end can often be achieved in a number of ways. An engineer has to resolve this set of choices to find the 'best' way. Like a medical doctor he gives professional opinions on what he considers the best course of action. The best way of doing something must include being the cheapest way. Remember that 'cheapest' must take into account running costs as well as capital costs. A machine which is cheap to buy may not turn out to be the best if running costs are high, and so running and capital costs must be low when taken together, though not so low that reliability is sacrificed. An engineer must therefore be a bit of an economist as well as a technologist. To quote A. M. Wellington in 1887, 'Engineering . . . is the art of doing that well with one dollar which any bungler can do with two, after a fashion'.

The work of an engineer is frequently misunderstood. Engineering is not all oily rags and fuse wire as some people imagine. The aspect of construction does, of course, come into it but it is the end product of a great deal of brainwork, research and development.

Engineers tend to be socially-minded people. They are usually concerned about the way in which their work helps others, and their job has a real social meaning and content.

What qualities go then to make a good engineer? It seems to me they are of two kinds, technical and social. For the first, he must understand the basic physics of the devices, apparatus and machines with which he works. He must be able to plan ahead, be cost-conscious and appreciate good engineering design. As for social skills the young engineer should be able to get on well with other people. Cooperation with others when they need a hand is the best way of ensuring help when you are in a tight spot.

The really promising engineers are the ones who are able to sift through various facts, arrive at the relevant and important ones quickly, and from these make the best decision. They must be able to justify and have faith in their decisions and be capable of convincing others of the correctness of their choice. It follows, therefore, that an executive engineer requires considerable managerial skills, i.e. he must be able to handle men as well as he handles materials.

If ever an electrical engineer needs a mnemonic to help him tackle a problem, let him recall his rivals, i.e. GAS, thus:

(a) *G*ather every relevant fact and factor.
(b) *A*nalyse the total problem.
(c) *S*olve the problem in the light of the facts using your technical knowledge.

Engineers have a tremendous advantage in being able to see the end product of their applied knowledge in tangible form.

Look around you — nearly everything has had an engineer's touch. Why not add yours to this? It is one of the most rewarding experiences to have.

Appendix I Types of Power Stations

Steam From Coal — The Boiler House

The illustration shows a modern boiler burning pulverized coal at rates up to 200 000 kg an hour. From the coal store, fuel is carried on a conveyor belt 1 and discharged by means of a coal tipper 2 into the bunker 3. It then falls, through a weigher 4, into the coal pulverizing mill 5, where it is ground to a powder as fine as flour. The mill usually consists of a round metal table on which large steel rollers or balls are positioned. The table revolves, forcing the coal under the rollers or balls which crush it.

Air is drawn from the top of the boiler house 6 by the forced air draught fan 7 and passes through the air preheaters 8 to the hot air duct 9. From here some of the air passes directly to the burners 10 and the remainder is taken through the primary air fan 11 to the pulverizing mill, where it is mixed with the powdered coal, blowing it along pipes to the burners 10 of the furnace 12. It mixes with the rest of the air and burns to produce a great amount of heat.

The boiler consists of a large number of tubes 13 extending to the full height of the structure, and the heat produced raises the temperature of the water circulating in them to create steam, which passes to the steam drum 14 at a very high pressure (possibly to 170 atmosphere). The steam is heated further in the superheater 15 and fed through the outlet valve 16 to the high pressure cylinder of the steam turbine 17. It may be hot enough to make the steam pipes glow a dull red (566 °C) and good thermal insulation must be used around them to prevent excessive heat loss.

When the steam has been through the first cylinder (high pressure) of the turbine, it is returned to the reheater of the boiler 18 and reheated before being passed through the other cylinders (intermediate and low pressure) of the turbine.

From the turbine the steam passes into a condenser 19 to be turned back into water called 'condensate'. This is pumped through feed heaters 20 (where it may be heated to about 250 °C) to the economizer 21 where the temperature is raised sufficiently for the condensate to be returned to the lower half of the steam drum 14 of the boiler. This re-use of water is necessary owing to the amount and standard of purity demanded by modern boilers.

The flue gases leaving the boiler are used to reheat the condensate in the economizer 21 and then pass through the air preheaters 8 to the electrostatic

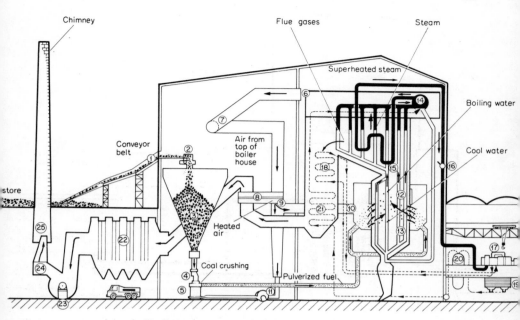

The boiler house. (By courtesy of C.E.G.B.)

precipitator 22. Finally, they are drawn by the induced draught fan 23 into the main flue 24 and to the chimney 25.

The electrostatic precipitator consists of metal plates which are charged with static electricity. Dust and grit in the flue gases are attracted on to these plates so that none passes up the chimney to pollute the atmosphere. Regular mechanical hammer blows cause accumulations of ash, dust and grit to fall to the bottom of the precipitator, where they collect in a hopper for disposal. Additional accumulations of ash also collect in hoppers beneath the furnace.

The ash is either sold for use in road or building constructions, or piped as a slurry of ash and water to a settling lagoon, where the water drains off. Once this lagoon (which may originally have been a worked out gravel pit) has been filled in, it can be returned to agricultural use or the ash can be removed for other purposes.

Oil-fired power stations have boilers that are very similar although there is no coal-handling and pulverizing plant.

Steam Raising — The Nuclear Reactor

The atom may be pictured as a central nucleus having a positive electric charge around which satellite electrons having a negative charge revolve. The nucleus is made up of protons, each having a positive charge, and neutrons which have no charge. Nuclear fission takes place when a free neutron strikes

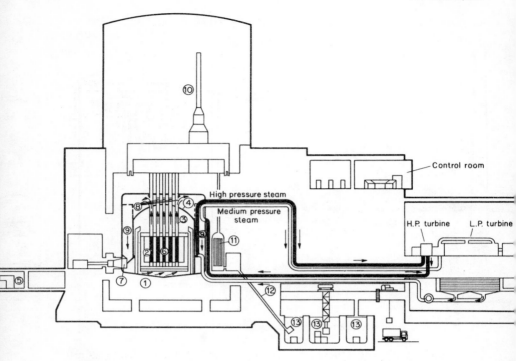

The nuclear reactor. (By courtesy of C.E.G.B.)

the nucleus of a fissile element, such as uranium 235. The nucleus splits into
two particles, releasing energy which appears as heat. Several new neutrons
are released by this splitting, some of these colliding with other fissile nuclei
which also split and so produce a chain reaction. Nuclear fission is most
likely when the neutrons are travelling slowly, and to achieve this the fissile
material is surrounded by a 'moderator' which slows them down. In a nuclear
reactor the complete process is controlled.

The illustration shows an Advanced Gas-cooled Reactor (A.G.R.) of the
type built for the second power station at Dungerness, Kent. The reinforced-
concrete, steel-lined pressure vessel 1 contains the reactor, boilers and ancil-
liary equipment. The reactor consists of a moderator 2 — a core of pure
graphite constructed from thousands of separate graphite blocks and containing numerous vertical channels 3 — surrounded by an inner pressure cylinder 4.

The enriched uranium dioxide fuel which is used in the reactor is kept in
a fuel store 5 until required. It is in pellet form and is sealed in stainless steel
cans forming a *fuel pin*. A cluster of 36 of these pins is arranged in a graphite
sleeve to form a *fuel element*. Eight of these fuel elements are tied together
to form a *fuel stringer*, one of which, 6, is placed in each vertical channel of

the core. Nuclear fission takes place in the fuel elements, and the heat liberated is carried away by streams of carbon dioxide gas under high pressure. The gas is pumped by gas circulators 7 up through the vertical channels in the moderator, passing around the fuel stringers and leaving by ports 8 to enter the boilers 9 at the top, where it heats water to produce steam.

Once the steam has been through the high pressure stage of the boiler, it is reheated in the boiler before passing through the remaining stages (as in a conventional power station). And again, after the steam has done its work in the turbine, it is passed into a condenser where it is condensed back to water and returned to the boiler.

A refuelling machine 10 is used to position and remove the fuel stringers in the reactor, and the fission process is controlled at a desired level of activity by moving control rods (not shown) containing boron-steel, a neutron-absorbing material, in or out of the graphite core. Used fuel stringers are removed from the reactor and stored in a special chamber 11 for about a week. They are then dismantled and the separate fuel elements are lowered down a chute 12 into a pond of water 13, where they remain until their radioactivity has decreased sufficiently for them to be removed from the power station.

Steam into Mechanical Power — The Turbine

From the boiler, a steam pipe 1 conveys steam to the turbine through a stop valve (which can be used to shut off steam in an emergency) and through control valves 2 that automatically regulate the supply of steam to the turbine. Stop valve and control valves are located in a steam chest and a governor 3, are driven from the main turbine shaft 4, operates the control valves to regulate the amount of steam used. (This depends upon the speed of the turbine and the amount of electricity required from the generator.)

Steam from the control valves enters the high pressure cylinder of the turbine, where it passes through a ring of stationary blades 5 fixed to the cylinder wall 6. These act as nozzles and direct steam on to a second ring of moving blades 7 mounted on a disc secured to the turbine shaft. The second ring turns the shaft as a result of the force of the steam. The stationary and moving blades together constitute a 'stage' of the turbine and in practice many stages are necessary, so that a cylinder contains a number of rings of stationary blades with rings of moving blades arranged between them. The steam passes through each stage in turn until it reaches the end of the high pressure cylinder, and in its passage some of its heat energy is converted into mechanical energy.

The steam leaving the high pressure cylinder goes back to the boiler for reheating, 8, and returns by a further pipe 9 to the intermediate pressure cylinder. Here it passes through another series of stationary and moving blades.

Finally the steam is taken to the low pressure cylinders, each of which it enters at the centre 10, flowing outwards in opposite directions through the rows of turbine blades — an arrangement known as double flow — to the extremities of the cylinder. As the steam gives up its heat energy to drive the

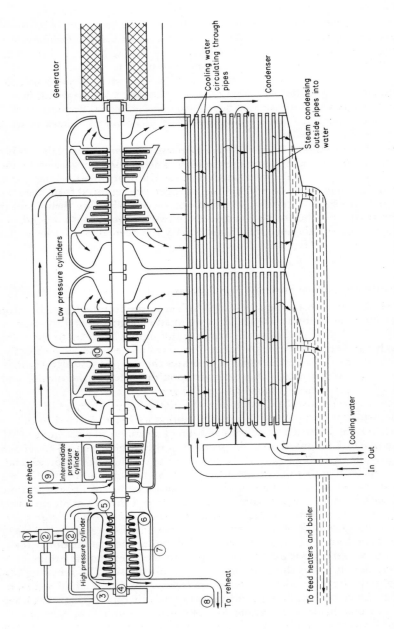

The steam turbine. (By courtesy of C.E.G.B.)

turbine, its temperature and pressure fall, and it is because of this expansion that the blades are wider and longer towards the low pressure end of the turbine.

When as much energy as possible has been extracted from the steam it is exhausted directly to the condenser. This runs the length of the low pressure part of the turbine and may be beneath or on either side of it. The condenser consists of a large vessel containing some 20 000 tubes each about three centimetres in diameter. Cold water from the river, estuary, sea or cooling tower is circulated through these tubes, and as the steam from the turbine passes round them it is rapidly condensed into water — condensate. Because water has a much smaller comparative volume than steam, a vacuum is created in the condenser. This allows the steam to be used down to pressures below that of the normal atmosphere and more energy can be utilized.

From the condenser, the condensate is pumped through low-pressure feed heaters by the extraction pump, after which its pressure is raised to boiler pressure by the boiler feed pump. It is passed through further feed heaters to the economizer and the boiler for reconversion into steam.

Steam Turbine Blades

There are two main types of blading arrangement used in steam turbines — (a) impulse and (b) reaction types. With an impulse turbine the high speed steam is directed on to 'buckets' whereas in the reaction turbine both stationary and moving blades have a similar shape. In a modern steam turbine the high pressure cylinders have impulse blading and the low pressure cylinders reaction blading. In many cases there is a mixture of the two types.

Hydro-electricity

Although the South of Scotland Electricity Board and North of Scotland Hydro-Electric Board possess many hydro-electric generating stations, the C.E.G.B. has only one big one at Ffestiniog, in Wales. The station is particularly interesting, however, because it is one of the few pumped water storage schemes. It uses a small natural lake, Llyn Stwlan, blocked across its natural outlet by a concrete buttress dam forming a large reservoir. The conduit system leading the water to a lower level consists of two vertical shafts, each of which divides into two tunnels having a fall of 1 in 40. The tunnels break out of the hillside above the generating station and are connected to it by four steel penstocks or water races. In falling from this height, the water gains considerable kinetic energy and is flowing very fast, so that when directed against the blades of the turbine it causes the turbine to rotate with great power. The water turbine has a vertical shaft and is mechanically coupled to a similarly aligned generator.

Water turbo-generators work at much lower speeds than steam turbo-generators but the electricity generated by them, is at the same frequency and is fed to the National Grid in the usual way. Once the water has passed through

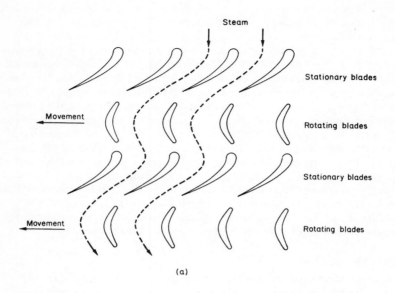

(a)

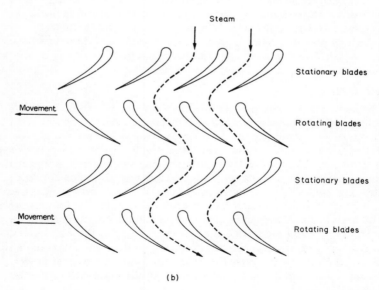

(b)

Steam turbine blades.
(a) Impulse turbine
(b) Reaction turbine.

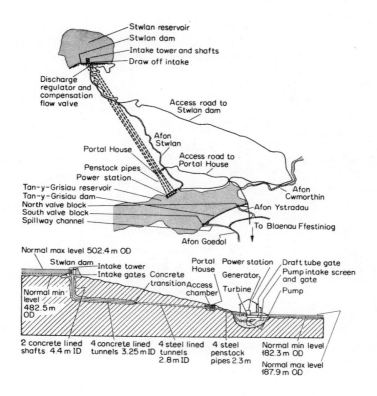

Ffestiniog Pumped Storage Scheme.
OD = ordnance datum, ID = inside diameter.
(By courtesy of G.E.C. Machines Ltd.)

the turbine runner and delivered up its energy it is carried along a 'tail race' to the lower reservoir. This lower reservoir has a much larger capacity than the upper one and during normal operation, its level varies by only a quarter of that of the upper one.

Water can be pumped up to the higher reservoir from the lower by using the generator powered to work as a motor. Pumping is arranged by closing the turbine inlet valve, engaging a mechanical coupling to connect the pump and opening the pump discharge valve. Whether pumping or generating, all the machines rotate in the same direction. When generating, the coupling is disengaged and the pump discharge valve closed, which isolates the storage pump completely. The total capital cost of the Ffestiniog pumped storage project, including transmission, is estimated at £15 million. This compares with an estimate of £25 million for an equivalent steam plant.

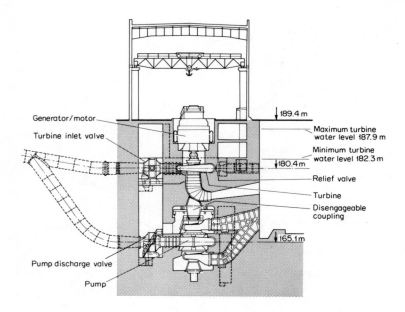

Cross-section through one of the Ffestiniog motor/generator-pump/turbine machines.
(By courtesy of G.E.C. Machines Ltd.)

Appendix II Appearance and Industrial Design

Most of the problems discussed have been largely technical and economic but they are not the only considerations. Electrical apparatus must not only do the job for which it is intended but it must also be efficiently designed in relation to manufacturing processes and materials, economic in production and in use and, important from the sales viewpoint, visually satisfying. Formerly some companies considered visual design as a luxury to be added at some later date in the production of an article, making it more acceptable to the intending purchaser. This superficial visual styling has very little to do with the fundamental design although it may add a glamorous overcoat.

The industrial designer is not concerned with 'beautifying' engineering products but with their form and function from their conception through all the processes of manufacture to the final product. Even after the strictly functional matters such as technical performance, efficiency and reliability have received the attention of engineers, there are still many factors to consider which are not written into the specification. A good designer is concerned with the exact way the product will be used, i.e. the man/machine relationship, and the product's appearance.

Consider the first of these two aspects, the man/machine relationship. Any product or equipment must be compatible with the limitations of the user. The designer's job, therefore, often involves the arrangement of the parts of the equipment and its controls, so that the operator can perform what is expected of him with the minimum of mistakes caused by fatigue. To do this the designer needs to study the exact way the product will be used. Relating the design of the equipment to the operator is one aspect of the science of ergonomics, which may be defined as planning the demands of the machine to the mental and physical capabilities of the operator. For brevity, we will concern ourselves with one feature only relating to power stations, that of controls.

Controls such as levers, knobs, switches, dials, etc., must always be ergonomically suited to the user and must be simple. In particular, switches ideally should indicate the direction and amount of operation (i.e. 1 turn, 180° turn, or 90° turn). Any layout should match a work study analysis of the job being done. The layout should also acknowledge *visual* balance and symmetry (not strictly physical symmetry) in order that the whole might be unified and harmonious.

167

Voltmeter face, old design.
(By courtesy of G.E.C. Ltd.)

Voltmeter face, new design.
(By courtesy of G.E.C. Ltd.)

Perhaps we are now in a position to discuss a few examples. Consider the simple example of the design of the front of a voltmeter which is to be placed on a control panel alongside many others. An operator might be expected to take readings of a number of instruments at quarter hour intervals, and at the same time keep a look-out for instruments indicating dangerous conditions. In this situation, the voltmeter face should give a clear indication of what is happening with the minimum of effort and possible errors on the part of the operator. Two possible designs are shown on page 168. Before reading on, decide which you feel is the better and why.

On the upper instrument, attention is first focused on the centre 'target' and next to the black outside case. On the lower instrument attention is focused first on the pointer and scale, the outside case being dark grey to avoid diverting attention. The different scale, numerals, and pointer on the lower instrument also make it more legible. Clearly, the lower instrument is superior. If the pointer is arranged to be vertical when on its normal reading during operation, any deviation from normal is immediately noticed.

Such design problems are not only confined to small apparatus such as voltmeters, but are found throughout the whole field of electrical power engineering.

Nowadays for example, it has become fashionable to fit a 'streamlined' case over the complete machine, supposedly to improve the external appearance of turbo-generators (see page 17). The machine looks efficient — if even a bit anonymous! Certainly, the case has tidied up the appearance and made everything neater, but perhaps amenity has been given too much priority over function. How can the operation engineers detect early signs of steam leaks and minor fires? By the time these have made themselves known by escaping through the case, it is difficult to contain the fault without a shut-down and possibly endangering the complete power station. Once steam or fire fills the inside of the case, you can't get inside to tackle the fault. Such things should not occur, but the fact is that they do! During the regular summer overhaul of plant, dismantling, storing, and re-erecting the case takes considerably longer than the actual maintenance of turbine and generator. Does the case warrant this expenditure of time and money? Good appearance should be an intrinsic part of the design, and there should be no need for a glamorous metal cover to be added afterward.

As an example of good ergonomic design, consider a power station control room. Here instruments, control switches, and mimic diagrams show the state of the power station. To enable the engineers to do their control and supervisory work under the most favourable conditions, the room is heated, sound-proofed, air-conditioned and well lit. From the shape of switch handles to the layout of instruments on panels, much thought has been given to obtaining the best results. But the ideal still has not been attained — the control room needs to be improved still further because of the very multiplicity of instruments, as shown by the photograph on page 170.

Starting a turbo-generator set after an overnight shut-down requires nearly 1000 operations to be performed in one hour from the conventional control room. How could this best be done? Was it reasonable to expect an engineer or group of engineers to work satisfactorily with this situation or should the whole process be completely automated? Unfortunately, to provide a fully

Ferrybridge 'C' power station. Part of the main control room at
the 2000 megawatt coal-fired Ferrybridge, Yorkshire, power
station. All four 500 megawatt boiler/turbine units, together
with auxiliary plant and gas turbines, are controlled from this
central area. The photograph shows the instrument panels and
control desks of two of the four units, each attended by an
operator and his assistant. On these early 500 megawatt units,
the size and number of control and switching devices led to
control desks 12 metre long, which are rather inconvenient
especially during start-up.
                         (By courtesy of C.E.G.B.)

automated system which would cope with all possible faults would be
extremely expensive (except in one or two experimental stations). The C.E.G.B.
thought that it would be reasonable to expect one man to be in charge of each
turbo-generator set and therefore compromized by devising a semi-automatic
system controlled by one man. A study of the timing and sequence of the
operations showed that by combining related operations the number of steps
could be reduced to fifty, with controls on a small desk within easy reach of
the seated engineer-in-charge (see photograph on page 171).

Cottam power station. One of the four unit panels in the central
control room at the 2000 megawatt Cottam, East Midlands, coal-
fired power station. Here, by careful planning and the use of
miniature controls and digital displays, it has been possible to
devise a control console on which all the controls needed for start-
up and subsequent operation are within arm's length of the operator. The photograph shows this console in the foreground.
Suspended above the console is an alarm panel housing the
relevant unit alarms arranged in urgent and non-urgent order,
all easily visible to the operator. Behind the console are
panels housing indicators and recorders for all the major
plant items which the operator has to survey.
(By courtesy of C.E.G.B.)

In this case the engineer was presented with a manageable number of
operations that required a degree of discretion or decision on his part. He
thus participated in start-up and was in a position to take prompt action
should a fault arise, and was not simply an observer. The participation aspect

was important since an engineer likes to feel in sympathy with his machine and this design philosophy has subsequently proved satisfactory.

Consider now the aspect of design concerned with the appearance of a product. This will be treated under three headings, *form*, *colour* and *texture*.

Primarily, the *form* of a product should display its function in relation to the user and the environment. The purpose and order of importance of the major components should be clearly stated and should not be hidden beneath any false facade of no functional value. Any cover should have tidy detailing.

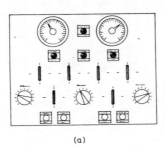

(a)

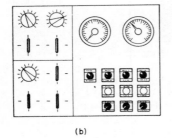

(b)

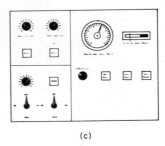

(c)

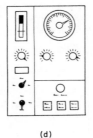

(d)

(e)

The method of lighting is important because strong directional light (e.g. sunlight) can visually alter form considerably, even producing dangerous shadows or highlights. Diffused artificial lighting is usually recommended for power station control rooms.

Form should be determined by the nature and purpose of the product. The materials should not be debased or tortured to something other than what they are. They should be used for the very virtues they possess.

Colour coding can be used to indicate job function. In a simple case, red push buttons are used for the stopping of a sequence and green push buttons for the starting, or safety to commence a sequence. In a more complex scheme the colour orange may be used for electrical switch boxes, blue for electric motors, yellow for circulating water-pumps and green for feed pumps, but unless this technique is used with restraint the total scheme may give a distressing kaleidescopic effect. A more subtle treatment may prove more effective. Colour can also be employed to indicate the order of assembly of parts during initial installation and subsequent maintenance.

On the plant itself light colours can help reduce apparent top heavy masses and darker colours reduce size and give stability lower down. In general, colour may be used to emphasize the virtue of the shape or form of the product or equipment.

Redesign of control panel layout (opposite).

Control panel layouts (a) often needlessly follow a symmetry that springs from a traditional regard for a balanced design.

A functional arrangement (b) however, not only assists the operator by grouping related components but can also simplify wiring.

A somewhat later and better design (c) replaces the ON/OFF switches with push buttons which light up when the circuit is alive. These push-buttons carry their own legend. The current indicator is an edgewise type and need not carry a scale although figures can be shown if necessary. The main point is to know if one is working in the 'safe area'. The knobs on the multiposition switches have been redesigned to give a clearer indication of position and a comfortable grip.

A new layout (d) uses the same components but in a smaller panel space.

This size (e) could be diminished still further by the use of intermediate electronic circuitry but it must be remembered that a machine-tool operator does not want minute knobs and push-buttons that one could use, for instance, in laboratory instruments.

(By courtesy of R. Kay Esq.)

Different psychological effects may be produced by different colours, e.g. warmth may be obtained with 'reds' and coldness with 'blues'. The former is assertive, the latter recessive. These considerations are important when creating working environments such as the power station control room.

Texture is that quality of a surface which should be pleasurable to touch, efficient in use and suitable to the object. A slight textural background is obviously good for working environments containing controls since high gloss finishes can produce undesirable highlights. In a room large areas of colour can be broken up or small areas unified using texture.

Appendix III Oil Break Circuit-Breaker

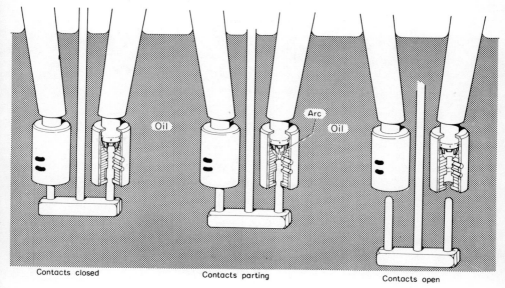

Contacts closed   Contacts parting   Contacts open

One phase of a 33 kV oil break circuit-breaker with two
NF-type interrupters, breaking an electric circuit.
(By courtesy of G.E.C. Switchgear Ltd.)

Description of interrupter unit

The diagram on page 176 shows a typical NF—natural flow—interrupter assembly, partly sectioned. This consists of a robust insulated cylinder case A which is attached to the contact top block B, and locked in the correct position by the locking tag. The contact block is fitted to the foot of the breaker bushing by means of clamp C.

    The interrupter case contains a series of 'labyrinths' D which are shaped to form 'passages' and 'blind alleys'. The 'passages' control the flow of oil and

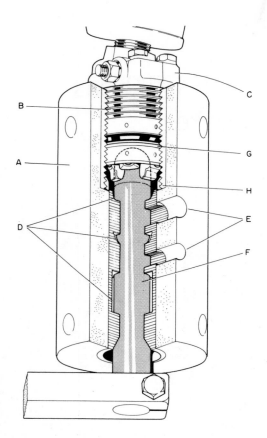

Typical NF-type interrupter Unit.
(By courtesy of G.E.C. Switchgear Ltd.)

oil gas created by the presence of an arc and the 'blind alleys' cause high pressure pockets to be set up during the same period.

On the side of the interrupter case there are a series of vents E, spaced in the correct position relative to the labyrinths.

When moving contact F is in the closed position, that is when it engages with the fixed contact assembly G, all the vents are blocked and during the opening period the amount of vent area exposed depends on the position of the moving contact F inside the interrupter.

The fixed contact assembly which comprises 'main' and 'arcing' fingers is completely detachable, and is held in position by the top labyrinth component H.

In the top block B there are a number of relief vents, the purpose of which is to control the maximum internal pressure in the interrupter during operation.

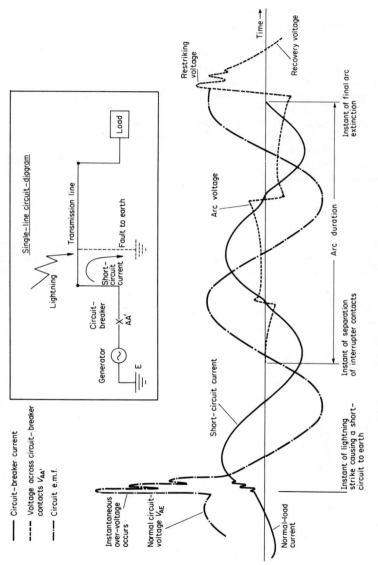

— Circuit–breaker current
--- Voltage across circuit–breaker contacts $V_{AA'}$
—·— Circuit e.m.f.

Single–line circuit–diagram

Generator

E

Circuit–breaker
X
AA'

Short–circuit current

Fault to earth

Lightning

Transmission line

Load

Restriking voltage

Recovery voltage

Time →

Arc voltage

Instant of final arc extinction

Arc duration

Instant of separation of interrupter contacts

Short–circuit current

Instantaneous over–voltage occurs

Normal circuit–voltage V_{AE}

Instant of lightning strike causing a short–circuit to earth

Normal–load current

Diagram illustrating the breaking of a short circuit current due to lightning by an oil circuit breaker.
(By courtesy of Reyrolle-Parsons Ltd.)

Operation

On opening an alternating current circuit under oil, an arc is created between the fixed and moving contacts vaporizing the oil in its vicinity. The oil gases generated are at a high pressure because of being confined to the restricted space in the region of the fixed contact chamber.

As the moving contact continues downward it unblocks the first side vent of the interrupter unit. Some of the high-pressure oil gas will escape but a simultaneous flow of oil shoots across the path of the arc before it can exhaust through the vent. The arc is therefore subjected to a jet of cool oil and at the same time forced into contact with the labyrinths. This will have the effect of vaporizing more oil thereby maintaining the internal pressure and at the same time cooling the arc by abstracting latent heat energy of vaporization. Thus the compression of the arc and its cooling are the two agents utilized to effect arc interruption and extinction.

At the times in the alternating current cycle when the current is near zero, the arc diminishes, momentarily 'pauses' and the action of the oil and gas become even more effective. Fresh oil and oil gas displace the 'conducting' atmosphere left by the arc and the dielectric strength of the contact gap increases.

The above sequence of events is repeated near every current zero, whilst the arc is being lengthened and progressively exposed to an increasing number of oil jets. In the final stage the breakdown voltage of the contact gap exceeds the system voltage and interruption is complete.

Appendix IV Air Blast Circuit-Breaker

There are various designs of air blast circuit-breaker but most up to 275 kilo-volt use the same principle of operation. This principle is to use a series of special switches called interrupters to initially break the circuit and once the current is stopped to isolate the input and output terminals by means of a make-switch. The number of interrupter units in series depends upon the voltage. The interrupter units are only open during the air blast and close automatically after it stops. The isolating arm does not break any circuit carrying current, although it is used to close a circuit. A typical breaker for 132 kilovolt is shown below.

To open the circuit breaker, the trip coil of an electro-pneumatic valve is energized. This opens to admit air from the receiver to the blast valve. The blast valve then opens and releases air which feeds, via a blast pipe, interrupters which all open simultaneously.

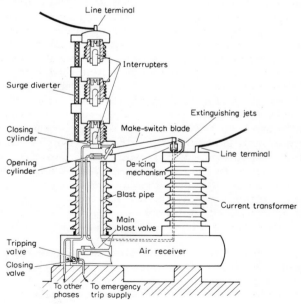

Typical air blast circuit-breaker.

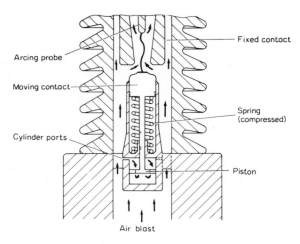

Typical air blast interrupter unit.

The operation of an interrupter is shown above. Under normal conditions the moving contact is held in a closed position by means of a spring. On release of the air, some of it passes into the top of the cylinder via the ports and forces the piston downwards, with assistance from the main blast of air which exerts a further downward pressure on the main contact. Damping is incorporated by the air trapped under the piston which is released through the small holes in the piston. The blast of air transfers the arc to the tip of the moving contact and the fixed probe, where it is extinguished.

After a 'built-in' interval of time, allowing a sufficient period for the arcs to be extinguished, a quantity of air is fed to the ice-breaking device which gives a sharp hammer blow to the tip of the make-switch blade to dislodge any ice that might have formed. At the same time air is passed to the opening cylinder which opens the make-switch blade. When the make-switch blade is fully open, it operates an auxiliary switch which de-energizes the trip coil.

The closing operation is much simpler than the opening operation since the interrupters remain closed. In this operation the closing-valve opens allowing air to pass from the receiver to the closing cylinder so that the make-switch blade closes.

An emergency trip device is incorporated which allows the circuit breaker to be tripped mechanically. Also, if one phase should fail to open, a back-up feature is incorporated which operates automatically.

To avoid condensation of moisture on insulating surfaces inside the circuit breaker, a continuous flow of dry air is passed through the circuit breaker.

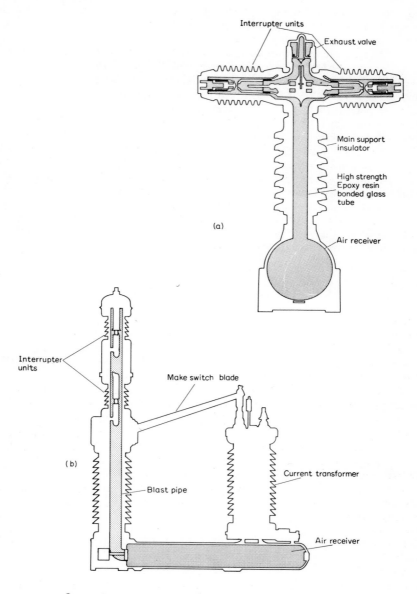

Comparison of permanently pressurized and non-permanently
pressurized air blast circuit-breakers.
(By courtesy of G.E.C. Switchgear Ltd.)

Surge diverters consisting of non-linear resistors are fitted across the interrupters to ensure an even voltage between interrupters and to limit the effect of overvoltages due to lightning and when switching.

For 275 kilovolt circuit-breakers a similar design is used with two units in series, joined together by male and female make-switch blades.

For the very high-voltage circuit-breakers, such as those used at 400 kilovolt, many manufacturers have brought out new designs employing interrupters which are permanently pressurized and having no make-switch. Circuit interruption is accomplished by switching the current first through parallel arcing contacts arranged in series with heavy resistors. The main switch contacts then open but no arc is formed. Shortly afterwards, the arcing contacts open and a blast of air attenuates the arc.

The advantage of the air blast circuit-breaker over the oil circuit breaker is that it has a slightly faster operating time and all the advantages to be gained by not using oil, i.e. no oil storage plant is required, there are no oil carbonizing problems and there is also a reduction in the fire risk. The disadvantages are that it has not got the mechanical simplicity of oil circuit breakers and it requires costly compressed air equipment.

Appendix V Automatic Control Systems

To illustrate the principle with a practical example, consider a bath of oil which is to be kept at a constant temperature by means of an electrical immersion heater. Assume that the bath temperature is to be measured with a simple mercury-in-glass thermometer and that the electric current for the immersion heater is supplied by a standard alternating current autotransformer or 'Variac'. By means of experiment, or otherwise, it is possible to determine the electric currents and the corresponding autotransformer dial-settings which give the different steady-state bath temperatures. Hence by turning the auto-transformer to the appropriate dial-setting, any current and hence any bath temperature can be slected. This is an open-loop control system. Unfortunately, the steady bath temperature also depends on ambient air temperature as well

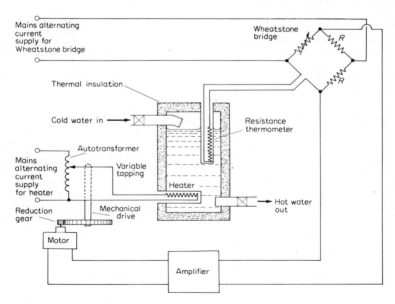

Automatic control of water temperature.

as on any fluctuations in the mains supply voltage. In other words, it is difficult to guarantee that a given dial-setting (input) on the autotransformer will give a current producing the correct bath temperature (output). The easiest but most undesirable method of overcoming such contingencies is to employ a human operator to act as temperature observer and make manual corrections to the dial settings as required. Such a system becomes a closed-loop control system since the human intervention has closed the loop joining output to input.

A more reliable closed-loop arrangement would be to eliminate the human operator and replace the link by some electrical or electromechanical system. Suppose the mercury thermometer is replaced by a platinum-wire resistance thermometer which is then fixed in one arm of a Wheatstone bridge balancing network (see diagram on page 183). (The bridge may be operated by a voltage derived from the mains but the balance, of course, is independent of supply voltage). The error voltage obtained from the bridge, as a result of a difference between required and actual temperature, is fed to a power amplifier, and hence to a motor driving the dial-settings on the autotransformer. The amplifier error voltage, drives the motor in such a way that the autotransformer dial-settings are increased or decreased to give the required correction.

Appendix VI Transducers

Although there are literally hundreds of transducer types, probably the most well-known is in the home — a record player pick-up in which a piezo-electric crystal or magnetic cartridge converts mechanical tracking variations on the record to electric currents which, after amplification, drive the loudspeaker. Three examples of transducers and their use should suffice to illustrate their importance in industry.

The tachometer is a small permanent magnet generator which produces an output voltage proportional to the speed of rotation of its armature. To measure the output speed of any motor, therefore, the tachogenerator armature is coupled directly to the output shaft of the motor. The voltage from the tachogenerator is proportional to the speed of the motor.

Thermocouple transducers are used in a boiler house to measure temperature, but the minute thermo-electric currents are usually insufficient to operate the necessary meters, recorders and relays, etc., and amplification is required.

Flame failure devices are common in oil-fired furnaces and domestic central heating systems. In such applications the transducer is a photocell which detects the light from the flame. Under normal conditions, with the flame on, the photocell current drives electronic equipment supplying a relay which holds the oil supply valve open. However, if the flame goes out the photocell cuts off and de-energizes the electronic equipment and relay thereby closing the oil valve. The oil injection into the combustion furnace then stops immediately.

Articles moving on a factory conveyor belt may be counted by interruption of a light beam focussed onto a photocell which is connected to some form of electronic or electro-mechanical counter via electronic amplifiers. By arranging light beams at different heights or in different positions, articles may be sorted for size, or alternatively, using the variation in current, they may be sorted for colour. For example, such electronic machines have been made for the sorting of fruits, seeds, nuts and similar foodstuffs into ripe or not-so-ripe categories according to their colour. In all these cases the photocell current drives amplifiers which power the mechanical handling equipment required to sort the articles onto different feeder conveyors. Alternatively, if articles interrupting the light beam were to be processed in some way, such as painting, the photocell current feeding the amplifier could actuate a switch or relay to turn on a spray gun.

Appendix VII Construction in Industrial Electronics

Voltage amplifiers and power amplifiers have been mentioned as important components in any control system. A short discussion of electronics hardware itself would seem, therefore, to be appropriate.

In its early years electronic equipment was essentially an assembly of individual components such as thermionic vacuum tubes, resistors, inductors and capacitors mounted on a metal chassis, connected together by means of insulated wires to form a complete circuit. Power supplies for the apparatus were usually direct current derived from either batteries or mains-driven rectifier units. Because of the nature of its construction, the equipment was relatively heavy and occupied a considerable space. Moreover, a good deal of skilled attention was needed in the repair of a fault, which was most often due to vacuum tube failure.

The 1939—45 War had important effects on the design of electronic equipment. Equipment had to withstand severe environmental conditions and be very reliable and readily maintained. In the later 1940s and during the 1950s many wartime innovations in the military field were introduced into industrial electronics. More complex circuits were developed and complete systems formed by interconnecting the various electronic stages. Modular methods of construction were employed, i.e. the equipment was built of 'electronic building bricks' which could be quickly located and replaced, and the repair executed at leisure. Miniature valves were introduced in the late 1940s and this led to a significant reduction in the size of electronic equipments.

A further big step forward was the introduction of printed wiring. The technique replaces all electrical insulated connecting wires with a circuit pattern printed in copper on an insulated board. The components are connected between appropriate points on the circuit pattern and soldered in position.

An important advantage of the method is that it is readily adaptable to mass production in a factory. Later, the printed wiring technique was supplemented by the printed circuit technique in which the passive components (resistors, inductors, capacitors, etc.) were also fabricated on to the insulating board. These components were achieved by the use of oxides, resistive inks and silver pastes. Printed wiring and printed circuits were not the only developments in the 1950s, for by this time the transistor had been invented, replacing the thermionic valve.

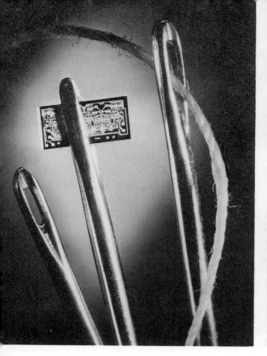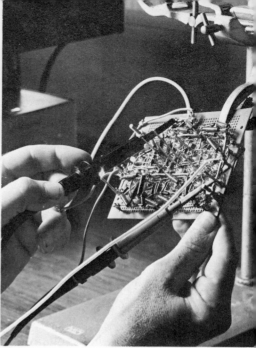

Manufacture of integrated circuits (above, left). The smallness of the silicon chip on which extremely complex circuits are made is illustrated by this photograph of an unencapsulated electronic circuit, called a decade counter, going through the eye of a no. 5 sewing needle; the circuit contains over 120 components, and the rope is ordinary 40 gauge sewing cotton.
(By courtesy of Mullard Ltd.)

Manufacture of integrated circuits (above, right). This is a 'bread board' of a high speed logic unit. In an integrated circuit form it is produced on a chip less than 2 square millimetre in area.
(By courtesy of Mullard Ltd.)

The most obvious differences between transistors and valves were their smaller size and more robust construction. Other advantages stemmed from their low operation voltage, high overall efficiency and the fact that they did not need heater or other special power supplies. Naturally the substitution of transistors for valves made possible the development of new products such as large computers which employed a very large number of electronic units.

Today we live in an age of electronic miniaturization; everything has been made smaller and smaller. Modern electronics is really micro-electronics. The basic units are called 'integrated circuits', i.e. circuits made as one entity and not as an assembly of separate components. Although there are many varieties in use today there are two basic types made by thin-film deposition or diffusion in semiconductors. The thin film circuit is usually built on to a ceramic or glass insulator by depositing various materials in selected areas by plating, evaporation, etc. The areas are controlled by masks placed on top of the insulating substrate when each material is laid down. The overall pattern and material structure produced by the separate masks is such that a complete network is achieved. The film thickness may range from 4—10 000 atoms. In the second method all the circuit components are fabricated into a single chip of semiconductor, usually silicon. The individual components such as resistors or transistors are obtained by diffusing different impurities in different concentrations at appropriate points in the chip. (Often a practical circuit of 'lumped' components is difficult to realize in this chip form and in such cases simulation techniques are employed.)

In both the thin film and semiconductor chip methods the final integrated circuit is encapsulated and hermetically sealed. The integrated circuit unit acts as an 'electronic building brick' having a particular function.

To give some idea of the size reduction which may be achieved, a complete amplifier employing ten transistors and associated resistors, capacitors and other components may be built into the size of a pin's head. This degree of miniaturization, although not so important in factory applications, is vital in the aeronautics and space industry. The real advantage of modern electronics to the industrialist is that it is not only more reliable but cheaper to buy and run than the old valve electronics.

Appendix VIII Digital Computers

Digital computers work with basic electronic units which may be 'on' corresponding to the digit one (1) or 'off' corresponding to the digit zero (0). The feature is important because electronic equipment works most reliably when switched in this mode — the effect of ageing and electronic interference or noise, etc., being minimized.

Because of these advantages it is convenient for computers to work with binary arithmetic based on two digits as distinct from the more conventional decimal arithmetic based on ten digits. Transfer between the the two systems is relatively easy so that for example the numbers 0, 1, 2, 3, 4, 5, 6, 7 in the decimal system correspond to 0, 1, 10, 11, 100, 101, 110 and 111 in the binary form. In practical cases information is usually fed into a computer in decimal form and then converted to binary form by a decimal to binary conversion. This information is then in a suitable form to be used by the computer in its calculations. Naturally, binary to decimal conversion is required at the output of the computer when the computation is completed so that the answer is more readily understandable.

Digital computers have two basic types of memory units which may be called temporary and permanent memory stores, respectively. The temporary store, accumulator or shift register units are made up of a cascade of elements called 'flip-flops' which may employ two discrete transistors and associated components, or a number of integrated circuits. The permanent memory store comprises a large number of basic units made up of magnetic ferrite cores, tapes or discs which can be polarized in one of two directions corresponding to the two states or binary digits. Because the basic units are so cheap and reliable, large memory banks can be built to form the computer.

Information on a programme in the form of instructions and data are fed into the computer by means of a set of punched cards or a roll of paper tape. Electromechanical sensing devices then translate the information into the machine language in a number of stages, including conversion, and the computations are performed. Calculated results are then reconverted from the machine language to the original language by means of other converters and printed out by a line printer.

Appendix IX Analogue Computers

The analogue computer is concerned with special sorts of equations called differential equations which are generally about dynamical behaviour, i.e. changes that occur with time. Variables occuring in the problems are represented on the computer by means of continuously varying voltages. The output, as a function of time, is examined as a continuous trace on a pen recorder, and the true behaviour of the actual output obtained by analogue with the actual trace. Analogue computers employ electronic units called operational amplifiers interconnected in various ways. Such amplifiers can perform the various mathematical functions of differentiation and integration. Any mathematical constants that may be involved in the solution are determined by the standard procedure of examining the initial conditions of the problem.

Appendix X Arago's Disc Experiment

Arago's results can be explained from Faraday's work in 1831 and Lenz's law first stated in 1834. Lenz's law states that when a current is induced in an electric circuit by a changing magnetic field, the induced current produces its own magnetic field, which opposes the original induction field. When the changing field is due to physical motion between conductor and field, the induced current is such as to oppose that motion.

The operation of the Arago disc can therefore be explained as follows.

The motion of the copper disc relative to the stationary compass needle induces a current in the disc. This current then produces its own magnetic field which drags the needle around. The only way for no relative motion to exist would be for the disc and compass needle to rotate in synchronism. This is impossible because there would be no induced current to cause the torque. Of necessity, therefore, there must be a difference of speed between compass needle and disc. This speed difference is called 'slip'. The amount of slip is just sufficient to ensure the disc currents generate the required torque.

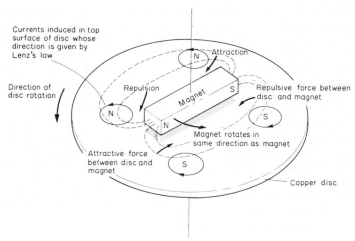

Explanation of Arago's experiment and the induction motor.

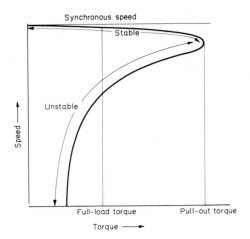

Induction motor speed/torque characteristics.

For example, a slip of 10 per cent with a disc rotating at 1000 revolutions per minute would mean a real speed difference of 100 revolutions per minute, giving an actual compass speed of 900 revolutions per minute.

In a commercial induction motor the magnetic field is caused to rotate by the action of alternating currents flowing in the stator winding. The speed of field rotation or synchronous speed depends on the mains frequency and number of stator poles. Set inside the stator core is a slotted rotor in which are embedded conductors short-circuited at their ends. When the motor is switched on, the rotating magnetic field 'cuts' the conductors in the stationary rotor and induces currents in them. Because of these currents and their associated magnetic fields, a torque is set up which turns the rotor in the same direction as the rotating field. If the rotor had to do no work it would catch up and rotate in synchronism with the rotating field. As, however, a certain amount of power is required to overcome friction, the actual rotor speed at the end of the initial acceleration period is always less than the synchronous speed. With light loads the percentage slip is small and the two speeds are nearly equal. With heavier loads the actual speed decreases and the slip increases to generate the rotor currents and give the extra torque. For standard machines the slip at full load is about 4 per cent so that for a 2-pole, 3000 revolutions per minute synchronous speed motor, the speed at full load is

$$3000 - (3000 \times \tfrac{4}{100}) = 2880 \text{ revolutions per minute}$$

For loads slightly greater than this the machine will stall (see diagram above).

Appendix XI Caption to Endpapers

At a coal or oil-fired power station the fuel is fed into a combustion furnace and the heat generated boils water in tubes to produce high pressure steam (see front end paper). In a nuclear power station the heat from the reactor is not used to produce steam directly but to heat carbon dioxide gas and then water, via a heat exchanger.

The steam is then piped to a steam turbine where it is directed on to the blades, causing the turbine shaft to rotate. After passing through the turbine the steam is at low pressure and is taken to a condenser to be cooled and converted back to water.

The turbine is mechanically connected to an electrical generator which consists of a fixed set of coils and a rotating electromagnet energized with current from a separate source. Alternating current of high power is generated in the fixed coils as the electromagnet rotates. The output current is delivered from the generator at three terminals. The combined turbine and generator is called a three-phase turbo-generator.

From the generator the electrical power is fed to a step-up transformer which raises the voltage to 400 000 or 275 000 volt. The power is then controlled by special switches en route to the high voltage transmission lines.

After transmission to the region where it is to be used the voltage is lowered by a step-down transformer at Bulk Supply Points to 132 000 volt. Subsequent voltage reduction may be obtained via an intermediate stage working at 66 000 to 33 000 volt. Electricity at 33 000 volt is at a level suitable for supplying power to towns and groups of villages. The 132 000, 66 000 and 33 000 volt lines form a *primary* distribution network to all consumers.

Outgoing lines at 33 000 volt radiate from the grid supply points and may directly feed big factories; alternatively, they may feed transformers in intermediate substations where the voltage is lowered further to 11 000 volt. Radial *secondary* distribution lines then radiate from these intermediate substations to small factories or 'local-area' distribution substations.

In the distribution substation, which may be at the centre of a housing estate, village, or shopping centre, the voltage is transformed to 415/240 volt; this voltage is more suitable for domestic and commercial use. Cables then carry the power down each street where it can be tapped off for the individual household or shop. Between the grid supply point and the distribution sub-station various switches at the different voltages direct and control the power flows.

The transmission and primary distribution system is arranged so that the interruption of a single line section does not disrupt the power supply to any particular area. This is achieved by connecting each of the generating stations, grid supply points and intermediate substations with a minimum of two links so that if one of them fails power can be supplied via an alternative route.

Further Reading

In the following references, *Electronics and Power, Journal of the Institution of Electrical Engineers,* is abbreviated to *I.E.E. (Electronics and Power),* and *Students' Quarterly Journal of the Institution of Electrical Engineers* is abbreviated to *I.E.E.S.Q.J.*

Chapter 1 Electricity Today

ANGELINI, A. M. 'Electricity supply in Italy', *I.E.E. (Electronics and Power),* Sept. 1967, p. 332.

BARNEA, J. 'Geothermal Power', *Scientific American,* Jan. 1972, p. 70.

BEAR, L. 'Turbovisory Gear', *I.E.E. S.Q.J.,* Sept. 1962, p. 12.

BENTHAM, F. 'Painting the stage – a review of British theatre lighting', *I.E.E. (Electronics and Power),* April 1966, p. 104.

BERRY, G. 'Street lighting today – a review of British practice', *I.E.E. (Electronics and Power),* Jan. 1967, p. 4.

BOARDMAN, F. D. 'Computing aids for the Grid control engineer', *I.E.E. (Electronics and Power),* Nov. 1971, p. 430.

BOOTH, E. S. 'Power Supply for 1970', *I.E.E. (Electronics and Power),* Nov. 1966, p. 385.

BOUD, J. 'Son et Lumiere 1963', *Electricity,* May/June 1963, p. 123.

BRAIKEVITCH, M. 'Pumped Storage and Ffestiniog', *Electrical Review,* 26 Feb. 1960.

BRISTOW, A. L. 'Road heating', *I.E.E. S.Q.J.,* Dec. 1969, p. 51.

BUCKINGHAM, G. S. 'World Fuel Policy', *Electricity,* Nov./Dec. 1965, p. 309.

CAMPBELL, R. H. 'Fast Reactors', *I.E.E. (Electronics and Power),* Oct. 1967, p. 373.

CARRIE, A. R. 'Tidal Power', *I.E.E. S.Q.J.,* Sept. 1957, p. 47.

CLARK, D. 'Electricity and Nuclear Power', *Electricity,* May/June 1967, p. 175.

CLARK, D. 'Power Station Siting', *Electricity,* March/April 1961, p. 92.

CLAYDON, J. B. 'Peace River hydroelectric project', *I.E.E. (Electronics and Power),* Oct. 1968, p. 397.

COATON, J. R. 'Innovations in Lighting', *I.E.E. (Electronics and Power),* Sept. 1971, p. 351.

CONEY, L. H. 'Partnership in Energy', *Electricity,* Nov./Dec. 1962, p. 295.

COOPER, D. R. 'The Snowy Mountains Project', *I.E.E. S.Q.J.,* Sept. 1957, p. 10.

CRISP, J. 'Electrical supplies for construction sites', *I.E.E. (Electronics and Power),* Jan. 1966, p. 20.

DEIGNAN, J. P. 'How about odour television?', *I.E.E. (Electronics and Power),* Jan. 1968, p. 15.

DENT, K. H. 'The development of gas-cooled reactors', *I.E.E. (Electronics and Power),* Oct. 1968, p. 366.

DICKENSON, G. C. 'Electricity in modern waterworks practice', *I.E.E. S.Q.J.,* Sept. 1962, p. 3.

DURRANT, H. 'Emergency Repairs', *Electricity,* May/June 1963, p. 129.

EAVES, P. S. K. 'Gas turbines for power generation', *I.E.E. (Electronics and Power),* July 1971, p. 257.

EDWARDS, R. 'Electricity and Economic Growth', *Electricity,* Sept./Oct. 1965, p. 273

EDWARDS, R. E. *Economic Planning & Electricity Forecasting.* (The Electricity Council 103/3 7500.10.66) Paper at World Power Conference, Tokyo Sectional Meeting, Oct. 1966.

EDWARDS, R. S. and BROWN, F. H. S. 'The replacement of obsolescent plant', *Electricity,* March/April 1962, p. 55.

EGAR, K. G. 'The New Fuel', *I.E.E. S.Q.J.*, Dec. 1961, p. 63.

ELECTRICITY COUNCIL, *Electricity Supply 1955–1965*. (The Electricity Council, 30, Millbank, London SW1.)

ENGLISH ELECTRIC, 'West Burton Power Station — The Turbines', *English Electric Journal*, Vol. 21, No. 6, p. 19.

EVANS, C. 'Electricity in the oil industry', *I.E.E. S.Q.J.*, Sept. 1963, p. 49.

FOREST, J. S. 'Making Electricity', *Faraday Lecture*, 1963–64 C.E.G.B. Publication G. 453, Sept. 1966.

FREETH, R. T. O. 'Office Lighting', *Electricity*, March/April 1961.

FRY, D. W. 'Nuclear Energy for power generation', *I.E.E. (Electronics and Power)*, April 1965, p. 124.

FULTON, A. A. 'The Cruachan pumped storage development', *I.E.E. (Electronics and Power)*, July 1966, p. 220.

GOULDING, E. W. 'Electricity for the world's peasant farmers', *I.E.E. (Electronics and Power)*, July 1965, p. 224.

GRAY, A. 'Electriculture' 1967 — Review of electricity in agriculture and horticulture', *Electricity*, Jan./Feb. 1967, p. 4.

GRAY, A. 'Pot Plant Production using 'tower' glasshouses', *Electricity*, Sept./Oct. 1965, p. 250.

GUSCOTT, W. J. 'The problems of electrical development in the traditional coal mining territories of the North East', *Electricity*, Nov./Dec. 1964, p. 321.

HARLEN, R. M. 'Electric Cars', *I.E.E. S.Q.J.*, Dec. 1968, p. 64.

HASLAM, R. J. 'Water-cooled reactors', *I.E.E. (Electronics and Power)*, Oct. 1967, p. 370.

HAWLEY, R. 'Magnetohydrodynamic power generation', *I.E.E. S.Q.J.*, Dec. 1960, p. 63.

HIME, M. W. 'Domestic Block Storage Heating', *Electricity*, Nov./Dec. 1962.

HOGERTON, J. F. 'The arrival of nuclear power', *Scientific American*, Feb. 1968, p. 21.

HOLMES, R. 'Common Ground — British archaeology and the electrical engineer', *I.E.E. (Electronics and Power)*, May 1967, p. 163.

HOSSLI, W. 'Steam Turbines', *Scientific American*, April 1969, pp. 100–111.

HUBBERT, M. K. 'The energy resources of the earth', *Scientific American*, Sept. 1971, p. 60.

HUDSON, W. 'The Snowy Mountains Scheme', *Electricity*, July/Aug, 1962. p. 180.

HUNT, D. F. 'The fifth foundation — cooking, water-heating, laundering, refrigeration and house heating', *Electricity*, May/June 1962, p. 119.

HURLEY, F. I. 'Nuclear Power Today', *I.E.E. (Electronics and Power)*, Oct. 1967, p. 363.

ILIFFE, C. E. 'The economics of nuclear power', *I.E.E. (Electronics and Power)*, Oct. 1967, p. 377.

JEREMY, K. W. C. 'Developing the Electric Car', *Electricity*, March/April 1966, p. 60.

KENNEDY, G. 'Power-Station Design — an outline of trends in the United States', *I.E.E. (Electronics and Power)*, Feb. 1966, p. 37.

KENT, G. E. 'The Broiler Industry', *Electricity*, Sept./Oct. 1961, p. 279.

LIGHTTOLLER, B. 'Let there be Light', *I.E.E. (Electronics and Power)*, Jan. 1969, p. 3.

LLEWELLYN, N. 'Commissioning and Initial Operation of a Modern Electricity Generating Station', *Electricity*, March/April 1964, p. 75.

MACKENZIE, J. R. and RITCHIE, P. W. 'The Electricity Supply Industry of France', *I.E.E. S.Q.J.*, March 1963, p. 123.

MACKENZIE, W. 'The English Electric generator/motor-turbine/pumps at Cruachan hydro-electric station', *English Electric Journal*, July/Aug. 1966, Vol. 21, No. 4.

Ibid., 'The English Electric generator/motor at Cruachan hydro-electric station', Sept./Oct. 1966, Vol. 21, No. 5.

MACKINEVEN, H. 'The Heart of the Highlands — Review of North of Scotland Hydro-Electric Board', *Electricity*, Nov./Dec. 1967, p. 398.

MARSH, N. F., 'The future of electricity in England and Wales', *Electricity*, Nov./Dec. 1964, p. 336.

MARSHAM, T. N. 'Will nuclear power pay off?', *I.E.E. (Electronics and Power)*, Feb. 1971, p. 57.

MCLEOD, T. S. 'Efficiency in Industrial Research and Development', *I.E.E.* (*Electronics and Power*), May 1968, p. 202.

MCNAB, I. R. 'Magnetoplasmadynamic Electric Power Generation', *I.E.E. S.Q.J.*, Sept. 1969, p. 13.

MELLING, C. T. 'Electricity and Prosperity', *Electricity*, July/Aug. 1963, p. 173.

MELLING, C. T. 'Long term planning in the electricity supply industry', *Electricity*, Jan./Feb. 1964, p. 3.

MELLING, C. T. 'The future of off-peak electricity', *Electricity*, Nov./Dec. 1965, p. 283.

MILNE, I. 'Off-Peak Storage Heating', *I.E.E. S.Q.J.*, Sept. 1967, p. 13.

NIXON, J. 'Electricity Beating Winter Building Bogey', *Electricity*, Jan./Feb. 1967, p. 11.

NUGENT, H. 'Trends in Domestic Lighting Fashion', *Electricity*, Sept./Oct. 1966, p. 270.

OUTRAM, W. J. 'Dounreay Prototype Fast Reactor', *I.E.E.* (*Electronics and Power*), Feb. 1970, p. 43.

PATERSON, H. 'The Dutch Hay Drying System', *Electricity*, July/Aug. 1967, p. 178.

PHILLIPS, R. H. 'The All-Electric New Town', *Electricity*, Nov./Dec. 1966, p. 295.

PLUCKNET, E. G. and STEWARD, S. F. 'Electricity's future in the home', *Electricity*, July/Aug. 1965, p. 172.

POLLARD, R. 'Electrical services in industralised buildings in Britain', *I.E.E.* (*Electronics & Power*), Feb. 1968, p. 65.

RAHIMTOOLA, A. F. 'Power development in West Pakistan', *I.E.E.* (*Electronics and Power*), Nov. 1965, p. 370.

RATCLIFF, G. 'Electric Cars', *Electricity*, May/June 1967, p. 210.

REECE, A. B. J. 'Power production from nuclear energy', *I.E.E. S.Q.J.*, Sept. 1954, p. 3.

REVELEY, P. V. 'Electricity for Sabah', *I.E.E.* (*Electronics and Power*), Jan. 1967, p. 16.

REYROLLE-PARSONS LTD., *Parsons Multi-Cylinder Turbines*, (Publication B25 Ref. C.1.67.)

RIPPON, E. C. 'Power Plants in the 1970's', *Electricity*, Nov./Dec. 1967, p. 450.

ROBERTSON, A. 'Balance of Power in Fuel', *Electricity*, March/April 1966, p. 75.

ROLLO, D. R. 'Electric power in Norway', *I.E.E.* (*Electronics and Power*), Sept. 1968, p. 371.

ROOT, W. K. 'Electric Heating in the U.S.A.', *I.E.E.* (*Electronics and Power*), March 1968, p. 97.

ROPER, J. F. 'Light for commercial buildings', *I.E.E.* (*Electronics and Power*), Aug. 1966, p. 297.

SCHOLEFIELD, S. 'Electricity in the Pottery Industry', *Electricity*, Nov./Dec. 1961, p. 342.

SCHUMAR, J. F. 'Reactor Fuel Elements', *Scientific American*, Feb. 1959, p. 37.

SCHURR, S. H. 'Energy', *Scientific American*, Sept. 1963, p. 111.

SEABORG, G. T. and BLOOM, J. L. 'Fast Breeder Reactors', *Scientific American*, Nov. 1970, p. 13.

SEARBY, P. J. 'Present worth evaluations as an aid to nuclear power decisions', *Atom 178*, Aug. 1971, p. 185.

SELF, H. and WATSON, E. M. *Electricity Supply in Great Britain*, (London: Allen & Unwin, 1952).

SHARPLES, J. T. 'Electric Road Heating'. *I.E.E.* (*Electronics and Power*), June 1964, p. 189.

SMITH, A. G. 'Floodlighting Church Buildings', *I.E.E.* (*Electronics and Power*), Nov. 1967, p. 417.

SMITH, C. C. 'Some Aspects of airport lighting', *I.E.E.* (*Electronics and Power*), March 1970, p. 85.

SOUTH OF SCOTLAND ELECTRICITY BOARD, *Hydro-electric power*, on application to Cathcart House, Glasgow S4.

SUMMERS, C. M. 'The conversion of energy', *Scientific American*, Sept. 1971, p. 148.

SUTHERLAND, P. 'The use of electrolytic cells to prevent marine infestation at coastal power stations', *I.E.E. S.Q.J.*, Sept. 1967, p. 52.

SYDNEY, P. 'The development of the storage heater', *Electricity*, Sept./Oct. 1962, p. 247.

SYKES, B. '£—£—£'s for Ash', *Electricity*, Sept./Oct. 1966, p. 248.

THRING, M. W. 'Fuel and power in the 21st century', *I.E.E.* (*Electronics and Power*), Jan. 1972, p. 3.

WAKEFORD, P. 'Electricity in Animal Environment', *Electricity*, Jan./Feb. 1963.

WALTERS, D. G. 'The applications & control of gas turbo-generators', *I.E.E. S.Q.J.*, 1965.

WARE, A. A. 'The quest for controlled thermo-nuclear power', *I.E.E.* (*Electronics and Power*), Jan. 1965, p. 12.

WEBB, K. G. 'The use of models in power station design', *Electricity*, Sept./Oct. 1965, p. 258.

WILCOCK, A. 'The World at Night', *Electricity*, March/April 1964, p. 70.

WILCOCK, A. 'Progress in Lighting', *Electricity*, Sept./Oct. 1963, p. 246.

WILCOCK, A. 'Hotel and Restaurant Lighting', *Electricity*, Nov./Dec. 1965, p. 315.

WILLIS, K. E. V. 'Energy Storage — an assessment', *I.E.E.* (*Electronics and Power*), Sept. 1965, p. 310.

WOODSON, R. D. 'Cooling Towers', *Scientific American*, May 1971, p. 70.

WOOLMER, R. 'Medical and biological applications of electrical engineering', *I.E.E.* (*Electronics and Power*), June 1964, p. 178.

Chapter 2 How Electricity is Generated

ALEXANDER, W. O. 'The competition of materials', *Scientific American*, Sept. 1967, p. 254.

BAX, H. 'The Economical Material for Aluminium Distribution Networks', *Aluminium 1970*, Vol. 46 (7), p. 475. Translation obtainable from Metal Information Services Ltd., or Aluminium Federation, 60 Calthorpe Road, Five Ways, Birmingham, B 15 1TN.

CARMICHAEL, I. F. 'Some thoughts on the synchronous machine', *I.E.E.* (*Electronics and Power*), March 1967, p. 90.

CARMICHAEL, I. F. 'The Synchronous Machine', *I.E.E. S.Q.J.*, March 1966, p. 127.

DAVIES, A. 'Aluminium for busbars', *I.E.E. S.Q.J.*, Sept. 1957, p. 54.

EASTON, V. 'Excitation of large turbo-

generators', *I.E.E.* (*Electronics and Power*), March 1964, p. 81.

EHRENREICH, H. 'The electrical properties of materials', *Scientific American*, Sept. 1967, p. 194.

GIBSON, J. 'Design and operation of large d.c. generators', *I.E.E. S.Q.J.*, Sept. 1965, p. 19.

GILLESPIE, B. D. 'Aluminium and the Electrical Industry', *I.E.E.* (*Electronics and Power*), June 1968, p. 231.

HAYWARD, B. F. W. 'Silicones in electrical engineering', *Electricity*, May/June 1962, p. 114.

HAZEL, J. R. 'Power station auxilliaries — a review of some practical aspects', *I.E.E.* (*Electronics and Power*), Dec. 1965, p. 430.

HODGSON, D. L. and PARK, G. A. 'Copper for the electrical industry', *I.E.E. S.Q.J.*, 1956, p. 101.

JOHNSON, H. 'The uses of aluminium in distribution systems', *E.R.A. Distribution Conference*, 3–6 Oct. 1967, Group 2 Paper 1.

KEFFER, F. 'The magnetic properties of materials', *Scientific American*, Sept. 1967, p. 222.

KELLY, A. 'The nature of composite materials', *Scientific American*, Sept. 1967, p. 161.

MILBURN, J. D. 'Skin and Proximity Effects in Heavy-Current Conductors', *I.E.E. S.Q.J.*, March 1969, p. 172.

NYHOLM, R. S. 'Metals and Intermetallic Bonding in Chemical Compounds', *B. A. Advancement of Sc.*, Vol. 23, No. 115, Jan. 1967, p. 438.

PEARSON, R. 'Plastics in engineering', *I.E.E. S.Q.J.*, March 1971, p. 223.

REISS, H. 'The chemical properties of materials', *Scientific American*, Sept. 1967, p. 210.

RICHARDSON, P. 'Stator Vibration in Large Two-Pole Generators', *Parsons Journal*, Summer, 1966.

RICHARDSON, P. 'Developments in Large Turbo Type Generators', *Parsons Journal*, Summer 1962.

ROBERTS, T. J. 'Heat flow in rotating electrical machines', *I.E.E. S.Q.J.*, June 1962, p. 183.

SMITH, C. 'The electrical supplies to the auxiliaries in a modern power station', *I.E.E. S.Q.J.*, Sept. 1967, p. 21.

STATHER, K. S. 'Cooling turbo-alternator windings', *I.E.E. S.Q.J.*, Dec. 1956, p. 62.

SUD, Y. D. 'Static and Dynamic Measurements in Large Generators', *I.E.E. S.Q.J.*, Dec. 1967, p. 79.

TURNER, M. J. B. 'Sliding electrical contacts', *I.E.E. S.Q.J.*, Dec. 1965, p. 67.

ZIMAN, J. 'The thermal properties of materials', *Scientific American*, Sept. 1967, p. 180.

Chapter 3 Transmission and Distribution

BANNER, E. H. W. 'Transmission and Supply Voltages', *I.E.E. (Electronics and Power)*, April 1969, p. 128.

BARLOW, H. E. M. 'Waveguides for power transmission and distribution', *I.E.E. (Electronics and Power)*, Nov. 1964, p. 391.

BARTHOLD, L. O. and PFEIFFER, H. G. 'High Voltage Transmission', *Scientific American*, May 1964, p. 38.

BATES, B. 'The problem of transformer noise', *I.E.E. S.Q.J.*, March 1966, p. 122.

BISHOP, K. W. J. and SPREADBURY, R. J. 'Static Inverters — recent developments and new applications', *I.E.E. (Electronics and Power)*, July 1968, p. 280.

BISHOP, W. M. 'Economics in the rural distribution of electricity', *I.E.E. (Electronics and Power)*, April 1968, p. 154.

BOLTON, B. 'The failure of electrical insulation by discharges', *I.E.E. S.Q.J.*, Dec. 1966, p. 77.

BOLTON, K. G. W. 'Electricity supply distribution — trimming for efficiency', *I.E.E. (Electronics and Power)*, April 1970, p. 124.

BOOTH, E. S.; CLARK, D.; EGGINGTON, J. L. and FOREST, J. S. 'The 400 kV Grid System for England and Wales', *Electricity*, May/June 1962.

BOX, R. W. G. 'Aviation in Electricity Supply', *Electricity*, May/June 1967.

BURDEN, B. T. 'High Voltage Transmission Practice', *I.E.E. S.Q.J.*, March 1965, p. 159.

CANNING, T. I. and BARRON, W. M. 'Lightning Measurement and Protection', *I.E.E. S.Q.J.*, March 1956, p. 181.

CASE, N. 'High Voltage Cable Practice', *I.E.E. S.Q.J.*, March 1956, p. 187.

C.E.G.B. *Why not underground?* C.E.G.B. Public Relations Branch, 15, Newgate St., EC1, Ref. G483 March 1967.

CHESTER, A. J. 'Electrical Protection', *I.E.E. S.Q.J.*, May 1957, p. 169.

COOKSON, A. H. 'Compressed gas as high voltage insulation', *I.E.E. S.Q.J.*, Dec. 1968, p. 105.

CRAGGS, P. A. 'Planning a works electrical distribution system', *I.E.E. S.Q.J.*, June 1969, p. 194.

DAVEY, V. S. 'Insulated Sodium Conductor — has it a future in Britain?' *I.E.E. (Electronics and Power)*, Nov. 1969, p. 395.

EDWARDS, R. S. and CLARK, D. 'Planning for expansion in electricity supply', *Electricity*, July/Aug. 1962, p. 184.

EDWARDS, R. 'The expansion of electricity supply', *Electricity*, July/Aug. 1963, p. 161.

EGGINTON, J. L. 'The Grid', *Electricity*, May/June 1961, p. 138.

ENGLISH ELECTRIC CO. LTD., *Proceedings of Conference on High Voltage Circuit Breakers. Their development and testing*, Sept. 1961, Stafford, U.K.

FAULKNER, M. G. 'Radio interference from transmission-line conductors', *I.E.E. S.Q.J.*, March 1971, p. 238.

FLURSHEIM, C. H. 'High voltage switchgear', *I.E.E. (Electronics and Power)*, Dec. 1971, p. 477.

FRANK, H. and FRISCH, I. T. 'Network Analysis', *Scientific American*, July 1970, p. 94.

GARRARD, C. J. O. 'Some recent developments in high voltage switchgear', *I.E.E. (Electronics and Power)*, June 1966, p. 194.

GILMORE, R. 'High voltage power transformer insulation', *I.E.E. S.Q.J.*, June 1967, p. 207.

GOLDHAMMER, J. 'New protection method for transmission towers', *I.E.E. (Electronics and Power)*, Sept. 1968, p. 375.

HARLEY, C. E. 'Cable Cooling Water Systems', *Electricity*, Jan./Feb. 1967.

HERBERT, J. R. 'Electricity and the Landscape', *Electricity*, Sept./Oct. 1967, p. 267.

HOLLINGSWORTH, D. T. 'Underground cables — past, present and future', *I.E.E. (Electronics and Power)*, Sept. 1969, p. 314.

LANE, F. J. 'D.C. Transmission tackles teething troubles', *The Engineer*, 1 Nov. 1968, p. 667.

LAWRENCE, R. 'Some factors affecting the electrical properties of transformer insulation', *I.E.E. S.Q.J.*, Sept. 1966, p. 44.

LEGG, D. 'The study of arc phenomena in air break circuit breakers', *I.E.E. S.Q.J.*, Sept. 1954, p. 20.

LIS, J. and THELWELL, M. J. 'Water cooling of e.h.v. cables', *I.E.E.* (*Electronics and Power*), June 1965, p. 210.

MATTHAIS, B. T. 'Superconductivity', *Scientific American*, Nov. 1957, p. 92.

MILNE, A. G. and MALTBY, J. H. 'Integrating the L.E.B. power system', *I.E.E.* (*Electronics and Power*), June 1967, p. 203.

MCANULLA, R. J. 'Sodium Conductor for Power Cables', *I.E.E.* (*Electronics and Power*), Nov. 1968, p. 434.

MEEK, J. M. 'The Vital Spark — lightning and spark breakdown in gases', *I.E.E.* (*Electronics and Power*), Nov. 1968, p. 431.

NORRIS, W. T. 'Superconductors for power engineering — today or tomorrow', *I.E.E. S.Q.J.*, Dec. 1970, p. 188.

PAUL, H. 'Power transmission of the future — microwaves or superconductors', *I.E.E.* (*Electronics and Power*), May 1970, p. 171.

PULSEFORD, H. E. and GUNNING, P. F. 'Power system control — the C.E.G.B.'s plans', *I.E.E.* (*Electronics and Power*), July 1967, p. 245.

RIEDER, W. 'Circuit Breakers', *Scientific American*, Jan. 1971, p. 76.

ROBBINS, J. A. 'Protecting the electricity user', *I.E.E.* (*Electronics and Power*), June 1971, p. 212.

SPARROW, A. C. 'The Cost of Undergrounding H.V. Transmission Lines', *Electricity*, Jan./Feb. 1965, p. 12.

SPURR, D. M. 'Hot Line Maintenance', *Electricity*, July/Aug. 1967, p. 183.

SYKES, J. H. M. 'High voltage d.c. power transmission — electronics on the largest scale', *I.E.E.* (*Electronics and Power*), July 1965, p. 243.

TAYLOR, F. W. 'Progress in High Voltage', *I.E.E.* (*Electronics and Power*), March 1966, p. 88.

TOWILL, S. 'Electricity Supply in Sweden', *I.E.E.* (*Electronics and Power*), Aug. 1965, p. 272.

TURNER, H. W. and TURNER, C. 'Choosing contact materials', *I.E.E.* (*Electronics and Power*), Nov. 1968, p. 437.

WHITNEY, W. B. 'The early history of the h.v. air-blast circuit breaker in the U.K. Part I'. *I.E.E.* (*Electronics and Power*), Jan. 1968, p. 17.

WHITNEY, W. B. 'The early history of the h.v. air-blast circuit breaker in the U.K. Part II', *I.E.E.* (*Electronics and Power*), Sept. 1968, p. 365.

WILLIAMS, A. L.; DAVEY, E. L.; and GIBSON, J. N. 'The Cook Strait d.c. power link', *I.E.E.* (*Electronics and Power*), Dec. 1965, p. 422.

WAY, A. N. 'Developments in high voltage direct current transmission', *I.E.E. S.Q.J.*, March 1965, p. 149.

Chapter 4 Utilization of Electrical Power

ACKROYD, M. H. 'Applications of Modern Control Theory', *I.E.E. S.Q.J.*, Dec. 1969, p. 68.

ALSTON, L. L. 'Electrical research in British Railways', *I.E.E.* (*Electronics and Power*), Jan. 1970, p. 3.

ARMSTRONG, D. S. 'Electro-magnetic pumps for liquid metals', *I.E.E. S.Q.J.*, Sept. 1962, p. 51.

ARNOLD, M. F. 'Thyristor drives for motors', *I.E.E. S.Q.J.*, June 1971, p. 271.

ASHBY, D. E. T. F. 'Electric propulsion in space', *I.E.E.* (*Electronics and Power*), Feb. 1968, p. 69.

BAGDZINSKI, I. 'Electrification of Railways', *S.Q.J. I.E.E.*, March 1956, p. 131.

BAGRIT, L. 'Automation — the Central Industrial Theme of the Sixties?', *Electricity*, July/Aug. 1962, p. 184.

BARNEY, G. C. 'The analysis of simple servomechanisms', *I.E.E. S.Q.J.*, Dec. 1962, p. 45.

BEADLE, R. E. 'Continuous Casting of Steel', *I.E.E. S.Q.J.*, Sept. 1963, p. 35.

BELCH, R.; ALEXANDER, C. E. and WINCH, K. 'Computer Control at Fawley', *I.E.E.* (*Electronics and Power*), May 1969, p. 152.

BENN, A. W. 'The social and political implications of automation', *I.E.E.* (*Electronics and Power*), June 1968, p. 243.

BENSON, F. A. 'The scope of modern electronics', *I.E.E.* (*Electronics and Power*), Jan. 1969, p. 11.

BONE, J. C. H. 'Motors in Industry', *Supplement to I.C.C.* (*Electronics and Power*), March 1972, p. 83.

BRINKER, C. D. 'Whither the Circuit Designer? — the impact of future trends in electronics', *I.E.E.* (*Electronics and Power*), March 1970, p. 110.

BROOK MOTORS LTD. *Installation and maintance of a.c. electric motors*, Publicity Department, Brook Motors Ltd., Empress Works, Huddersfield.

BROUGHALL, J. A. 'Electric traction in railways of the future', *I.E.E.* (*Electronics and Power*), Feb. 1967, p. 46.

BUNCE, L. C. 'Recent developments in the utilisation of electricity in the manufacture and processing of high grade steels', *Electricity*, Sept./Oct. 1962, p. 234.

CARLISLE, S. S. 'Automation and Industry', *I.E.E.* (*Electronics and Power*), Nov. 1966, p. 383.

COBBE, B. M. 'Traffic Control for West London', *I.E.E.* (*Electronics and Power*), April 1967, p. 118.

COLLIE, A. A. 'Solid state control of domestic appliances', *I.E.E.* (*Electronics and Power*), Jan. 1972, p. 19.

COOKSLEY, C. G. 'Electricity in the Iron & Steel Industry', *I.E.E. S.Q.J.*, June 1957, p. 193.

COONS, S. A. 'The uses of computers in technology', *Scientific American*, Sept. 1966, p. 176.

COWIE, D. G. 'Railway Electrification at Industrial Frequency', *I.E.E. S.Q.J.*, Sept. 1957, p. 38

CRABB, J. 'Computer memory systems', *I.E.E. S.Q.J.*, June 1971, p. 247.

CRIPPS, L. G. 'Electronics in cars', *I.E.E.* (*Electronics and Power*), Nov. 1970, p. 394.

CROSS, D. N. 'Industrial Frequency Electric Motive Power', *I.E.E. S.Q.J.*, Sept. 1957, p. 43.

DOBIE, W. C. 'Electricity for Industry', *Electricity*, Sept./Oct. 1966, p. 256.

DUMMER, G. W. A. 'Integrated electronics — an historical introduction', *I.E.E.* (*Electronics and Power*), March 1967, p. 71.

E.N.E.A. and the Norwegian Inst. for Atomic Energy, 'On-line computer for nuclear reactors', *Nuclear Engineering International*, Nov. 1968, p. 950.

EVANS, D. C. 'Computer Logic and Memory', *Scientific American*, Sept. 1966, p. 74.

FANO, R. M. and CORBATÓ, F. J., 'Time Sharing and Computers', *Scientific American American*, Sept. 1966, p. 128.

FIRNBERG, D. 'Information, management and computers', *I.E.E.* (*Electronics and Power*), Sept. 1971, p. 355.

GALLAGHER, L. V. and OLD, B. S. 'The continuous casting of steel', *Scientific American*, Dec. 1963, p. 74.

GEBALLE, T. H. 'New superconductors', *Scientific American*, Nov. 1971, p. 22.

GIBSON, H. J. 'Electro-Heat Studies', *Electricity*, July/Aug. 1962, p. 166.

GOODYEAR, C. J. 'Computers in Control', *I.E.E. S.Q.J.*, Dec. 1971, p. 319.

GREENBERGER, M. 'The uses of computers in organisations', *Scientific American*, Sept. 1966, p. 192.

HAMILTON, W. F. and NANCE, D. K. 'Systems Analysis of Urban Transportation', *Scientific American*, July 1969, p. 19.

HAYDEN, J. T. 'Superconductivity and its possible applications in transportation', *I.E.E.* (*Electronics and Power*), Aug. 1969, p. 281.

HEATH, F. G. 'Large-scale integration in Electronics', *Scientific American*, Feb. 1970, p. 22.

HENDER, B. S. 'Future of the battery-electric car', *I.E.E.* (*Electronics and Power*), Aug. 1964, p. 250.

HILL, H. 'Concorde electrics', *I.E.E.* (*Electronics and Power*), Oct. 1971, p. 402.

HITTINGER, W. C. and SPARKS, M. 'Microelectrics', *Scientific American*, Nov. 1965, p. 57.

HOFFMAN, G. A. 'The Electric Automobile', *Scientific American*, Oct. 1966, p. 34.

INSTITUTION OF ELECTRICAL ENGINEERS 'Modern industrial control', *Special issue of I.E.E.* (*Electronics and Power*), Nov. 1971.

JOHNSON, P. E. 'The programming of digital computers', *I.E.E. S.Q.J.*, March 1968, p. 168.

JONES, J. T. and WILLIAMS N. J.

'Computer control of steel works production', *I.E.E.* (*Electronics and Power*), June 1964, p. 182.

KING, G. 'Electric Circuits — past, present and future', *I.E.E.* (*Electronics and Power*), Sept. 1965, p. 299.

KNIGHT, C. E. 'The Application of Electronic Accounting Procedures in Area Boards', *Electricity*, May/June 1964, p. 135.

LAITHWAITE, E. R. 'Linear induction motors for high speed vehicles', *I.E.E.* (*Electronics and Power*), July 1969, p. 230.

LAITHWAITE, E. R. and BARWELL, F. T. 'Linear Induction Motors for High Speed Railways', *I.E.E.* (*Electronics and Power*), April 1964, p. 100.

LAVER, F. J. M. 'Stimulus or constraint? — the interplay of computer design and use', *I.E.E.* (*Electronics and Power*), Sept. 1968, p. 361.

LAWRENSON, P. J. 'Linear Electrical Machines', *I.E.E. S.Q.J.*, March 1960, p. 92.

LEVY, S. L. 'Taming a technology: some aspects of the microcircuit industry', *I.E.E.* (*Electronics and Power*), Sept. 1969, p. 320.

LIPETZ, B. A. 'Information Storage and Retrieval', *Scientific American*, Sept. 1966, p. 224.

MACE, D. G. W. 'Electricity in the service of road-traffic control', *I.E.E.* (*Electronics and Power*), April 1964, p. 120.

MASON, B. J. 'Forecasting weather by computer', *I.E.E.* (*Electronics and Power*), Jan. 1968, p. 4.

MAYR, O. 'The Origins of Feedback Control', *Scientific American*, Oct. 1970, p. 110.

McCARTHY, J. 'Information and Computers', *Scientific American*, Sept. 1966, p. 64.

McNAB, I. R. and WILKIN, G. A. 'Carbon fibre brushes for super-conducting machines', *I.E.E.* (*Electronics and Power*), Jan. 1972, p. 8.

MINSKY, M. L. 'Artificial Intelligence', *Scientific American*, Sept. 1966, p. 246.

MORRIS, G. A. and CHRISTIAN, P. L. 'Impressions of computers and uses', *I.E.E. S.Q.J.*, March 1970, p. 100.

MORTLOCK, J. R. 'Computers and Engineers', *I.E.E.* (*Electronics and Power*), Oct. 1966, p. 351.

NICOL, C. T. 'Electricity in Paper-Making', *Electricity*, Nov./Dec. 1965, p. 293.

NIX, G. F. 'Reciprocating Electrical Machines', *I.E.E. S.Q.J.*, June 1964, p. 222.

OETTINGER, A. G. 'The uses of computers in science', *Scientific American*, Sept. 1966, p. 160.

OSBORNE, A. K. *An Encyclopaedia of the Iron and Steel Industry* (London: The Technical Press, 1956.).

PADWICK, G. C. 'Integrated-circuit applications — a look to the future', *I.E.E.* (*Electronics and Power*), March 1967, p. 78.

PARKER, J. H. 'A look inside the thyristor', *I.E.E. S.Q.J.*, Sept. 1970, p. 147.

PARKINSON, T. E. 'Automation in Urban Transport', *I.E.E.* (*Electronics and Power*), Oct. 1968, p. 392.

PARTRIDGE, A. N. 'Electrical equipment for motor vehicles — recent developments', *I.E.E.* (*Electronics and Power*), Nov. 1970, p. 400.

PIERCE, J. R. 'The transmission of computer data', *Scientific American*, Sept. 1966, p. 144.

PINDER, R. S. 'Thin-Film Microelectronics', *I.E.E. S.Q.J.*, Sept. 1970, p. 153.

PIPPARD, A. B. 'Superconductors in industry', *I.E.E.* (*Electronics and Power*), June 1964, p. 195.

PLAYFAIR, E. 'The computer and the use we make of it', *I.E.E.* (*Electronics and Power*), June 1964, p. 205.

RAJCHMAN, J. A. 'Integrated Computer Memories', *Scientific American*, July 1967, p. 18.

RICHARDSON, A. 'Automation in Map Making', *I.E.E.* (*Electronics and Power*), Dec. 1968, p. 477.

RICHEY, G. 'Guided missiles a review of their history and design features', *I.E.E. S.Q.J.*, March 1968, p. 139.

RIDDLESTONE, H. G. 'Electrical equipment for flammable atmospheres — are our standards good enough', *I.E.E.* (*Electronics and Power*), Nov. 1967, p. 409.

ROBBINS, R. M. 'The Victoria Line and its successors', *I.E.E.* (*Electronics and Power*), June 1970, p. 226.

ROBERTS, A. D. 'Continuous Casting', *A.E.I. Engineering, Metal Industries Supplement*, p. 37. U.D.C. 62 1.746.047.

ROBERTSON, F. A. 'Packaging techniques

for modern microelectronics equipment', *I.E.E.* (*Electronics and Power*), Oct. 1971, p. 406.

ROBINSON, L. M. 'Telecommunications support for the Apollo programme', *I.E.E.* (*Electronics and Power*), July 1970, p. 244.

ROSSER, J. A. 'The development of fifty-cycle traction', *I.E.E. S.Q.J.*, December 1966, p. 68.

RUTHERFORD, D. 'Hybrid Computers', *I.E.E. S.Q.J.*, Dec. 1968, p. 68.

SAVAGE, R. 'Numerical control of machine tools', *I.E.E. S.Q.J.*, June 1967, p. 231.

SCOTT, A. W. 'Digital Computers and the Engineer', *I.E.E. S.Q.J.*, Sept. 1970, p. 162.

SCOTT-KERR, R. J. 'Computer-aided control at Pembroke power station', *I.E.E.* (*Electronics and Power*), March 1972, p. 87.

SHEPHERD, R. 'A computer-controlled flying shear', *I.E.E. S.Q.J.*, March 1964, p. 143.

STEVENSON, P. D. 'The Thyristor', *Electricity*, Nov./Dec. 1966, p. 326.

STEWART, G. F. 'The Automatic Factory', *I.E.E. S.Q.J.*, 1956, p. 111.

STRACHEY, C. 'System Analysis and Programming', *Scientific American*, Sept. 1966, p. 112.

SUPPES, P. 'The use of computers in education', *Scientific American*, 1966, p. 206.

SUTHERLAND, I. E. 'Computer Inputs and Outputs', *Scientific American*, Sept. 1966, p. 86.

THOMPSON, L. H. 'The evolution of electronic data processing and its effect on Area Board Accounting', *Electricity*, March/April, 1964, p. 82.

THRING, M. W. 'Automation in the home', *I.E.E.* (*Electronics and Power*), Nov. 1968, p. 440.

TRUSCOTT, D. N. 'Computers in control of processes', *I.E.E.* (*Electronics and Power*), June 1964, p. 199.

WALKER, R. I. and NAYLOR, R. 'Microelectronics technology — a look at fabrication methods', *I.E.E.* (*Electronics and Power*), March 1967, p. 75.

WARNER, M. G. R. 'Driverless farm tractors', *I.E.E.* (*Electronics and Power*), Aug. 1971, p. 308.

WESTBROOK, M. H. 'Electronic measurement in the automobile industry', *I.E.E.* (*Electronics and Power*), Nov. 1970, p. 406.

WILLIAMS, J. C. 'Picture of an induction motor', *I.E.E. S.Q.J.*, Dec. 1966, p. 90.

WILLIAMS, J. G. E. 'Electricity in paper manufacture', *I.E.E. S.Q.J.*, Sept. 1963, p. 17.

WILLIAMS, P. 'The Stockholm Underground Railway', *I.E.E. S.Q.J.*, March 1969, p. 133.

YEOMANS, K. A. 'Ward Leonard drives — 75 years of development', *I.E.E.* (*Electronics and Power*), April 1968, p. 144.

Chapters 5—8 The History of Electrical Engineering

ADBY, C. A. 'The early years of the power transformer', *I.E.E.* (*Electronics and Power*), Aug. 1968, p. 335.

APPLEYARD, R. *Charles Parsons*, (Constable, 1933).

BARAK, M. 'Georges Leclanché (1839—82)', *I.E.E.* (*Electronics and Power*), June 1966, p. 184.

BELLAMY, D. 'Taken for granted—Part V— Twenty four hour service', *Electricity*, July/Aug. 1962, p. 163.

BLACK, S. 'Sebastian de Ferranti — a man of vision', *Electricity*, March/April 1964, p. 91.

BLAKE, L. R. 'The rise and fall of public transport systems 1066—2066', *I.E.E.* (*Electronics and Power*), Aug. 1968, p. 326.

BOWERS, B. 'Wheatstone and the generation of electricity', *I.E.E.* (*Electronics and Power*), July 1970, p. 253.

BOWERS, B. *R.E.B. Crompton-Pioneer Electrical Engineer. A Science Museum Booklet*, H.M.S.O.

BROWN, S. 'A century of development in the supply industry', *I.E.E.* (*Electronics and Power*), April/May 1971, p. 149.

DERRY, T. K., and WILLIAMS, T. I. *A short history of technology*, Chapter 22, p. 680 (The Electrical Industry).

DUNSHEATH, P. *A history of electrical engineering*, (London: 1962) Faber & Faber.

GALLON, W. A. 'Taken for granted—Part VI—

Wayleaves', *Electricity*, Sept./Oct. 1962, p. 243.

GREEN, G. N. 'Taken for granted—Part X—Control of Area Board Distribution Systems', *Electricity*, May/June 1963, p. 116.

HALACSY, A. A. 'Anyos Jedlik, an inventor of the dynamo electric principle', *I.E.E.* (*Electronics and Power*), Sept. 1971, p. 332.

HAMMOND, B. J. 'Two thousand years of therapeutic electricity', *I.E.E.* (*Electronics and Power*), June 1969, p. 190.

HILLS, S. M. 'The Development of the Electric Lamp', *Electricity*, May/June 1965, p. 122.

HONEY, P. 'The history of electric lighting Part 1', *Electricity*, May/June 1966, p. 144.

HONEY, P. 'The history of electric lighting Part 2', *Electricity*, July/Aug. 1966, p. 182.

HONEY, P. 'The history of electric lighting Part 3', *Electricity*, Sept./Oct. 1966, p. 252.

HONEY, P. 'Electric Traction in Britain Part 1', *Electricity*, Nov./Dec. 1966, p. 320.

HONEY, P. 'Electric Traction in Britain Part 2', *Electricity*, Jan./Feb. 1967, p. 29.

HONEY, P. 'Electric Traction in Britain Part 3', *Electricity*, March/April 1967, p. 101.

HONEY, P. 'Electric Traction in Britain Part 4', *Electricity*, May/June 1967, p. 212.

HORSLEY, W. D. 'A scientific survey of North East England — Electrical Engineering', *British Association for the Advancement of Science*, 1949, p. 142.

HOWARD, P. R. 'Electricity's Research Effort — the British Picture', *I.E.E.* (*Electronics and Power*), Aug. 1967, p. 298.

INSTITUTION OF ELECTRICAL ENGINEERS — Teeside Sub-Centre Booklet, *From Telegraphs to Nuclear Power — a history of electrical engineering in Cleveland, South Durham, North Eastern Centre Jubilee 1919—1969.*

IRVING, D. B. 'Taken for granted—Part IV—The Billing of Accounts', *Electricity*, May/June 1962, p. 107.

JARVIS, C. M. 'The rise of electrical science', *J.I.E.E.*, Jan. 1955, p.13.

JARVIS, C. M. 'Towards the new light', *J.I.E.E.*, March. 1955, p. 145

JARVIS, C. M. 'Machinery for the new light Part I', *J.I.E.E.*, May 1955, p. 280.

JARVIS, C. M. 'Machinery for the new light Part II', *J.I.E.E.*, Sept. 1955, p. 566.

JARVIS, C. M. 'The origin and development of the electric telegraph Part I', *J.I.E.E.*, March 1956, p. 130.

JARVIS, C. M. 'The origin and development of the electric telegraph Part II', *J.I.E.E.*, p. 584.

JARVIS, C. M. 'The development of the electrical load', *J.I.E.E.*, June 1957, p. 310.

JARVIS, C. M. 'The evolution of the central station', *J.I.E.E.*, June 1958, p. 298.

JARVIS, C. M. 'Some Victorian electricians', *I.E.E.* (*Electronics and Power*), April 1966, p. 121.

JARVIS, C. M. 'Nikola Tesla and the Induction Motor', *J.I.E.E.*, Dec. 1970, p. 436.

JARVIS, C. M. 'The generation of electricity', Chapter 9 p. 177. *A history of technology* Vol. 5 (1850—1900). Editors: Singer, C.; Holmyard, E. J.; Hall, A. R.; Williams, T. I. (Oxford: Clarendon Press).

JARVIS, C. M. 'The distribution and utilisation of electricity', in *A history of technology*, Vol. 5 (1850—1900), Editors: Singer, C.; Holmyard, E.J.; Hall, A. R.; Williams, T. I. (Oxford: Clarendon Press), Chapter 10, p. 208.

JOFFE, A. F. 'The Revival of Thermo-electricity', *Scientific American*, Nov. 1958, p. 31.

JONES, D. A. 'Research in the electronics industry', *I.E.E.* (*Electronics and Power*), March 1965, p. 92.

KING, R. 'The Supply Industry's Perspectives', *Electricity*, Nov./Dec. 1961.

LAY, R. K. 'The history and changing fortunes of the inductor alternator', *I.E.E.* (*Electronics and Power*), Dec. 1968, p. 485.

MARSH, N. 'Taken for granted—Part I—Continuity of Supply on an extensive mains system'. *Electricity*, Nov./Dec. 1961, p. 338.

MATSCHOSS, C. *Great Engineers.*
Werner von Siemens p. 261
Charles Algernon Parsons p. 304
Thomas Alva Edison p. 314
George Westinghouse p. 333

MCSHANE, I. E. 'The Evolution of the Electrical Machine', *I.E.E. S.Q.J.*, June 1969, p. 229.

PARSONS, R. H. 'The early days of the power station industry', (Cambridge: University Press, 1939).

PARSONS, R. H. *The development of the steam turbine*, (Constable: 1936).

PULSEFORD, H. E. 'Taken for granted— Part XII—Continuity of electricity supply at low cost', *Electricity*, Sept./Oct. 1963, p. 233.

ORCHARD, F. A. 'A history of a pioneer undertaking', *J.I.E.E.*, 1939, p. 49.

RIDDING, A. S. Z. *de Ferranti — Pioneer of electrical power*, A Science Museum Booklet, H.M.S.O.

RIX, S. M. 'Taken for granted—Part VII— The work of the legal department', *Electricity*, Nov./Dec. 1962, p. 303.

SHARLIN, H. I. *The Making of the Electrical Age*, Abelard-Schuman, 1963.

SHARLIN, H. I. 'From Faraday to the dynamo', *Scientific American*, May 1961, p. 107.

SHIERS, G. 'The Induction Coil', *Scientific American*, May 1971, p. 81.

SIEMENS, E. *Inventor and Entrepreneur — recollections of Werner von Siemens*, (London: Lund Humphries, 1966).

SMITH, R. H. 'The London Electrical Engineers', *I.E.E. (Electronics and Power)*, June 1968, p. 239.

STEWART, G. N. 'Electricity, Supply in the South of Scotland', *Electricity*, Jan./ Feb. 1967, p. 39.

WELBOURN, D. B. 'Two industrial revolutions — or three?', *I.E.E. (Electronics and Power)*, Aug. 1969, p. 284.

Chapter 9 Epilogue

BRAYSHAW, G. S. 'Mathematics and the education of electrical engineers', *I.E.E. (Electronics and Power)*, June 1968, p. 235.

BROWN, S. 'The fascination of electrical power engineering', *I.E.E. (Electronics and Power)*, Nov. 1967, p. 407.

COOKSON, A. H. 'Council of Engineering Institutions', *I.E.E. S.Q.J.*, June 1967, p. 196.

DANIEL, F. W. 'Approaches to Economy and Productivity — relating to the electrical industry', *I.E.E. (Electronics and Power)*, Jan. 1968, p. 21.

DRAPER, D. W. 'The Application Engineer', *I.E.E. S.Q.J.*, June 1966, p. 223.

EDMUNDSON, D. 'Electrical Manufacture — today and tomorrow', *I.E.E. (Electronics and Power)*, Nov. 1969, p. 392.

EDWARDS, R. 'Human Values and Human Problems in Electricity Supply', *Electricity*, Sept./Oct. 1963, p. 223.

ENDERSBY, J. C. 'Can engineers manage?', *I.E.E. (Electronics and Power)*, Oct. 1971, p. 400.

FAIRBROTHER, E. 'What is an engineer?', *I.E.E. (Electronics and Power)*, June 1971, p. 219.

FANNING, P. J. 'The role of the electrical engineer in modern society', *I.E.E. S.Q.J.*, Sept. 1966, p. 41.

FERRANTI, S. Z. 'Electrical Engineering— today and tomorrow', *I.E.E. (Electronics and Power)*, April/May 1971, p. 169.

FREEMAN, C. 'Industrial innovation: the key to success', *I.E.E. (Electronics and Power)*, Aug. 1971, p. 297.

GARRATT, A. 'Value — the criterion of design', *I.E.E. (Electronics and Power)*, July 1968, p. 285.

HARRAP, G. V. 'Management and the engineer', *I.E.E. S.Q.J.*, Sept. 1971, p. 297.

HARRISON, R. J. 'Engineering Arts', *I.E.E. (Electronics and Power)*, July 1969, p. 243.

HOYLE, R. 'Engineers and Scientists — understanding the differences', *I.E.E. (Electronics and Power)*, Feb. 1967, p. 49.

HUMPHREYS, O. W. 'Science and Engineering', *I.E.E. (Electronics and Power)*, Nov. 1964, p. 376.

I.T.B. (Industrial Training Board). Booklet No. 5. The Training of Professional Engineers.

I.T.B. (Industrial Training Board). Booklet No. 9. The Training of Technician Engineers.

INGLIS, S. 'What makes Women Engineers?', *I.E.E. S.Q.J.*, March 1970, p. 83.

JAMES, B. M. 'Human Factors in Technology', *I.E.E. S.Q.J.*, March 1970, p. 86.

LEE, A. W. 'The design process — the birth of a new product', *I.E.E. (Electronics and Power)*, Nov. 1968, p. 442.

MCFADZEAN, W. 'The British Electrical Industry in the World Scene', *Electricity*, July/Aug. 1962, p. 177.

MELLING, C. T. 'The Engineer as Manager', *Electricity*, Nov./Dec. 1962, p. 293.
MITCHELL, J. W. E. 'Science and human values', *I.E.E.* (*Electronics and Power*), Jan. 1968, p. 2.
MORTON, H. B. 'The engineer and the law', *I.E.E.* (*Electronics and Power*), June 1969, p. 205.
SAY, M. G. 'Electrical and Electronic Engineering — a professional career', *I.E.E. S.Q.J.*, June 1966, p. 195.
SHEPHERD, W. G. 'Education for the Engineering Mission', *I.E.E.* (*Electronics and Power*), April 1968, p. 157.
SNOW, C. P. 'The place of the engineer in society', *I.E.E.* (*Electronics and Power*), May 1966, p. 152.
STEWARD, S. 'Britains electrical industry — its progress, problems and prospects', *I.E.E.* (*Electronics and Power*), March 1972, p. 80.
TAYLOR, J. 'The role of the consulting engineer', *I.E.E. S.Q.J.*, June 1971, p. 261.
THRING, M. W. 'The social responsibility of the engineer', *I.E.E.* (*Electronics and Power*), Aug. 1967 p. 292.

Appendix II

CRAFT, P. C. R. 'Ergonomics — the human factor in engineering', *English Electric Journal*, Vol. 21. No. 6, p. 6.
KAY, R. M. S. *Industrial Design in Engineering, Engineering Materials & Design*, Feb. 1959, and *Metropolitan-Vickers Gazette*, Sept. 1959.
KAY, R. M. 'Industrial Design in Electrical Engineering', *A.E.I. Engineering*, July/Aug. 1965, p. 202 UDC 7.05 : 621.3.
MATCHETT, E. 'Creative design — a new approach to fundamentals', *I.E.E.* (*Electronics and Power*), Dec. 1966, p. 436.

Appendix VII

DIXON, R. G. 'Ever decreasing circles — a history of electronics', *I.E.E. S.Q.J.*, March 1967, p. 131.

Index